HOMOLOGY AND DYNAMICAL SYSTEMS

Conference Board of the Mathematical Sciences

REGIONAL CONFERENCE SERIES IN MATHEMATICS

supported by the
National Science Foundation

Number 49

HOMOLOGY AND DYNAMICAL SYSTEMS

by

JOHN M. FRANKS

Published for the
Conference Board of the Mathematical Sciences
by the
American Mathematical Society
Providence, Rhode Island

Expository Lectures
from the CBMS Regional Conference
held at Emory University
August 18–22, 1980

Research supported in part by NSF Grants MCS 7701080 and MCS 8002177.

1980 *Mathematics Subject Classification*. Primary 58F09, 58F10, 58F11, 58F12, 58F13, 58F14, 58F15, 58F20, 58F25; Secondary 57R50, 57R65, 57R80.

Library of Congress Cataloging in Publication Data

Franks, John M., 1943–
    Homology and dynamical systems.
    (Regional conference series in mathematics, ISSN 0160-7642; no. 49)
    "Expository lectures from the CBMS regional conference held at Emory University, August 18–22, 1980"–Verso t.p.
    Bibliography: p.
    1. Differentiable dynamical systems. 2. Homology theory. I. Conference Board of the Mathematical Sciences. II. Title. III. Series.
QA1.R33 no. 49 [QA614.8]   510s [515.3′52]   82-8897
ISBN 0-8218-1700-0

Contents

Preface

The qualitative study of the solutions to ordinary differential equations has had a long and varied history. In recent years much attention has been paid to the connections between the theory of these smooth "dynamical systems" and two other areas of mathematics: ergodic theory and statistical mechanics on the one hand and algebraic and differential topology on the other.

It is the relationships between dynamical systems and topology which these lectures address. This particular part of dynamical systems dates from the work of Poincaré, especially his beautiful theorem equating the Euler characteristic of a surface with the sum of the indices of rest points of a flow on the surface.

In this century the most important contributions to this area of investigation have been made by Marston Morse and Steve Smale. It is not possible to survey all their contributions in a meaningful way in this brief introduction. However, the importance of their contributions can perhaps be gauged by the number of times their names occur in the chapter titles of these lectures. It would be remiss however not to mention the special importance of Smale's paper *Differentiable dynamical Systems* [S1], not only for investigations of the type we consider here, but all qualitative investigation of smooth dynamical systems. In addition to proving important new results, this article had a major influence on the direction of the whole field of dynamical systems. Two influences merit special mention. First it emphasized classifying dynamical systems according to the complexity of their qualitative dynamical behavior, rather than, for example, the more traditional way of classification by complexity of the algebraic form of the differential equation. Secondly Smale drew attention to structurally stable systems as particularly worthy of investigation and conjectured a characterization of them, which has subsequently been proven correct in many cases.

The relationship between qualitative dynamics and topology is much too large an area to consider fruitfully in its entirety within the framework of these lectures and accordingly we will narrow our attention to a collection of results with a particularly homological flavor.

The theme of these lectures is illustrated in the following diagram.

$$\underline{\text{Dynamics}} \xleftarrow{\substack{\text{Differential} \\ \text{topology}}} \substack{\text{Chain complex} \\ \text{description}} \xleftrightarrow{\text{algebra}} \underline{\text{Homology}}$$

(basic sets
and unstable
manifolds)

The left-hand arrow indicates that techniques of differential topology provide a connection between unstable manifolds (which we will see often form a cell decomposition) and a homological description of the system at the chain complex level. The right-hand arrow represents many algebraic results relating chain complex behavior with corresponding homological behavior. The bidirectionality of the arrows reflects the need, given one set of data, to understand what are the possibilities for the corresponding data at the other end of the arrow. The ultimate aim, of course, is to answer questions like, "In a given homological configuration, what kinds of dynamics can occur?" or "Given a dynamical configuration what are its homological implications?" This underlying theme recurs throughout these notes in many different guises, applied to many different classes of flows and diffeomorphisms.

In choosing the topics of these lectures compatibility with this theme has been my first criterion. In addition I have tried not to overlap too much with previous sets of lectures [B2, C, Gu, Mar, New, Sh], all of which deal with dynamical systems. This consideration and the unfortunate realization that I cannot include everything has led me to omit, for example, any discussion of the entropy conjecture, which has an unquestionable right to be in any treatise on homology and dynamical systems. In defense of its omission here I can only refer the reader to Chapter 5 of [B2].

Many conversations with colleagues too numerous to mention have been invaluable during the preparation of these lectures. Special thanks, however, are due to Steve Batterson, who is responsible for the existence of these lectures, as well as some of the theorems in them. I also wish to thank George Francis for drawing the illustrations to Appendix B.

Chapter 1. Hyperbolic chain recurrent sets

The object of these lectures is to study qualitatively the time evolution of smooth dynamical systems. We will limit our attention to smooth flows on compact manifolds and their discrete time analogues, smooth diffeomorphisms. In fact, we will place further limitations on the systems considered and the aim of this chapter is to describe and motivate these restrictions.

A qualitative analysis of the time evolution of a dynamical system should include a description of the long run or asymptotic behavior of points of the manifold M under the system, i.e., a description of the behavior of $f_t x$ as t tends to $\pm\infty$ (or $f^n x$ as $n \longrightarrow \pm\infty$ for a diffeomorphism). Ideally we should have this description for all $x \in M$.

It is clear that in such a description a special role is played by the periodic orbits and stationary points for flows and the periodic points for a diffeomorphism. It turns out that the set of points satisfying a kind of recurrence weaker than periodicity is crucial to the description of long run behavior.

(1.1) DEFINITION. If $f: M \longrightarrow M$ is a diffeomorphism, then $x \in M$ is called *chain recurrent* if for any $\epsilon > 0$ there exist points $x_1 = x, x_2, \ldots, x_{n-1}, x_n = x$ (n depends on ϵ) such that $d(fx_i, x_{i+1}) < \epsilon$ for $1 \leqslant i < n$, where $d(\ ,\ )$ is a metric on M. For a flow f_t, $x \in M$ is chain recurrent if for any $\epsilon > 0$ there exist points $x_1 = x, x_2, \ldots, x_n = x$ and real numbers $t(i) \geqslant 1$ such that $d(f_{t(i)}x_i, x_{i+1}) < \epsilon$, for $1 \leqslant i < n$. In either case the set of chain recurrent points is called the *chain recurrent set* and will be denoted by R or $R(f)$.

It is an easy exercise to show that the chain recurrent set R is invariant under the flow or diffeomorphism and is closed and hence compact since we are assuming M to be compact. One might think of R as the points which come within ϵ of being periodic for every $\epsilon > 0$.

The importance of chain recurrence for the description of asymptotic behavior of orbits of the system is shown in the following theorem of Conley.

(1.2) THEOREM [C]. *If f_t is a continuous flow on M, there exists a continuous function $g: M \longrightarrow R$ such that*
   (1) *If $x \notin R(f_t), g(f_t x) < g(f_s x)$ when $t > s$.*
   (2) *If $x, y \in R(f_t)$ then $g(x) = g(y)$ if and only if for $\epsilon > 0$ there exist points $x_1 = x, x_2, \ldots, x_n = y, x_{n+1}, \ldots, x_{2n} = x$ in R and real numbers $t(i) > 0, 1 \leqslant i < 2n$, such that $d(f_{t(i)}x_i, x_{i+1}) < \epsilon, 1 \leqslant i < 2n$.*

The obvious analogue of this theorem for diffeomorphisms is also valid.

(1.3) DEFINITION. A function $g: M \longrightarrow R$ satisfying the conclusion of this theorem (or its analogue for diffeomorphisms) is called a *Lyapunov function*.

Results of W. Wilson [W] show that it is possible to choose a Lyapunov function g which is C^∞ and satisfies $(d/dt)(g(f_t x))_{t=0} < 0$ for $x \notin R$. The existence of Lyapunov functions will be very important for much of our investigation. It is already clear from property (2) that R decomposes into invariant "chain transitive" pieces and from (1) that, for any $x \in M$, $f_t x$ tends to one of these pieces as $t \longrightarrow \infty$ (and to another as $t \longrightarrow -\infty$). Thus a qualitative analysis of a flow or diffeomorphism breaks naturally into two parts:

I. A topological description of the system restricted to the chain recurrent set R, and how points not in R approach R as time tends to ∞ or $-\infty$.

II. An analysis of how the sets of points tending toward the various pieces of R fit together to form the manifold M, and in particular how the topology of M is related to the behavior of the system restricted to R.

It is primarily this second aspect of the analysis which these lectures address. The class of dynamical systems we shall consider is a large subset of all smooth systems, but one whose elements possess rather strong stability properties. Since it is precisely this stability which has motivated the attention accorded to these flows and diffeomorphisms, we define it before proceeding.

(1.4) DEFINITION. Two diffeomorphisms $f, g: M \longrightarrow M$ are said to be *topologically conjugate* provided there is a homeomorphism $h: M \longrightarrow M$ such that $f = h^{-1} \circ g \circ h$. Two flows f_t and g_t on M are called *topologically equivalent* if there is a homeomorphism $h: M \longrightarrow M$ which carries orbits of f_t homeomorphically onto orbits of g_t and preserves the orientations given to the orbits by the flows.

(1.5) DEFINITION. A diffeomorphism $f: M \longrightarrow M$ is called *structurally stable* provided there is a neighborhood N of f in the space of diffeomorphisms of M with the C^1 topology, such that each $g \in N$ is topologically conjugate to f.

A flow f_t is called *structurally stable* provided there is a neighborhood N of the vector field generating f_t in the space of C^1 vector fields on M such that each flow g_t generated by a vector field in N is topologically equivalent to f_t.

The philosophy behind the attention given to structurally stable systems is that any dynamical system abstracted from nature and modeling a repeatable phenomenon can only be known approximately. Moreover the model must necessarily assume many influences on the system to be constant, while in reality they are subject to minor variation. Thus if repeated experimentation gives rise to the same qualitative behavior for some phenomenon it is plausible to assume that the "true" system governing the phenomenon is stable in the sense that any perturbation of it will exhibit essentially the same qualitative behavior. The concept of structural stability, introduced by Andronov and Stepanov, formalizes this sameness of qualitative behavior. It requires any perturbation of a structurally stable system to have an orbit structure homeomorphic to the original. A limitation of this line of reasoning worth bearing in mind is that the nature of some systems makes it plausible to consider only a restricted set of perturbations (e.g. those preserving a symmetry of the system or conserving energy) and these systems, while quite important, seldom have any stability properties of the type just described.

A great deal of research has been devoted to the problem of characterizing those systems which are structurally stable. While conditions which are sufficient for structural stability of a diffeomorphism or flow are known, it is unknown if these conditions are in general also necessary. The first of these conditions is that the chain recurrent set R possess a hyperbolic structure, which we now define.

(1.6) DEFINITION. A compact invariant set Λ for a diffeomorphism $f: M \rightarrow M$ is said to have a *hyperbolic structure* provided the tangent bundle of M restricted to Λ can be written as a Whitney sum $T_\Lambda M = E^u \oplus E^s$ of subbundles invariant under the derivative Df, and that there are constants $C > 0$, $\lambda \in (0, 1)$, such that

$$\|Df^n(v)\| \leqslant C\lambda^n\|v\| \quad \text{for all } v \in E^s, n \geqslant 0$$

and

$$\|Df^n(v)\| \geqslant C^{-1}\lambda^{-n}\|v\| \quad \text{for all } v \in E^u, n \geqslant 0.$$

If Λ is a compact invariant set for a flow it is said to have a *hyperbolic structure* provided that $T_\Lambda M$ is the Whitney sum of three bundles $E^u \oplus E^s \oplus E^c$ each invariant under Df_t for all t, that E^c is spanned by the vector field generating f_t, and that there are constants C, $a > 0$ such that

$$\|Df_t(v)\| \leqslant Ce^{-at}\|v\| \quad \text{for } v \in E^s, t \geqslant 0,$$

and

$$\|Df_t(v)\| \geqslant C^{-1}e^{at}\|v\| \quad \text{for } v \in E^u, t \geqslant 0.$$

If Λ is a stationary point for the flow f_t then E^c is taken to be vacuous.

The importance of hyperbolicity for stability properties of dynamical systems was observed by Smale in his seminal article [S1]. The property of having a hyperbolic chain recurrent set is equivalent to what Smale called *Axiom A and the no-cycle property* (see [F-S]). In particular the periodic orbits of the system are dense in R whenever R is hyperbolic, so in this case we can give an alternate description of R as the closure of periodic orbits of the system.

An important consequence of hyperbolicity of R is that the number of pieces into which R decomposes is finite. The following result of Smale is from [S1].

(1.7) SPECTRAL DECOMPOSITION THEOREM. *Suppose that the chain recurrent set R of a flow or diffeomorphism has a hyperbolic structure. Then R is a finite disjoint union of compact invariant sets $\Lambda_1, \Lambda_2, \ldots, \Lambda_n$ and each Λ_i contains an orbit of the system which is dense in Λ_i.*

These compact invariant sets with dense orbits are called *basic sets*. For a flow, basic sets can also be described as connected components of R. Since any basic set Λ contains a dense orbit and the vector bundles E^u and E^s are invariant, it follows that the restriction of

E^u and E^s to Λ must have fibers of constant dimension. The fiber dimension of the bundle E_Λ^u is called the *index* of Λ.

The second property of structurally stable systems concerns the stable and unstable manifolds of points in the chain recurrent set. The following result on stable manifolds for hyperbolic sets is from [H-P].

(1.8) STABLE MANIFOLD THEOREM. *Suppose Λ is a compact invariant hyperbolic set for a diffeomorphism f. Then for each $x \in \Lambda$ the sets*

$$W^s(x) = \{y \in M | d(f^n v, f^n x) \to 0 \text{ as } n \to \infty\}$$

and

$$W^u(x) = \{y \in M | d(f^{-n} y, f^{-n} x) \to 0 \text{ as } n \to \infty\}$$

are each the image of a smooth injective immersion of a Euclidean space into M. Moreover the tangent space $T_x(W^s(x)) = E_x^s$ and $T_x(W^u(x)) = E_x^u$. The manifolds $W^s(x)$ and $W^u(x)$ are called respectively the stable and unstable manifolds of x.

The analogous result for flows with $W^s(x) = \{y \in M | d(f_t x, f_t y) \to 0 \text{ as } t \to \infty\}$ and $W^u(x) = \{y \in M | d(f_t x, f_t y) \to 0 \text{ as } t \to -\infty\}$ is also valid.

For a flow we can also define the *weak stable manifold* $w^s(x)$ and the *weak unstable manifold* $w^u(x)$ by

$$w^s(x) = \bigcup_{t=-\infty}^{\infty} W^s(f_t(x)) \quad \text{and} \quad w^u(x) = \bigcup_{t=-\infty}^{\infty} W^u(f_t(x)).$$

Each of these is an injectively immersed copy of the bundle E^s or E^u restricted to the orbit of x, i.e. if the dimension of E_x^s is k then $w^s(x)$ is either

(a) an immersed k-plane, if x is a stationary point,

(b) an immersed $(k + 1)$-plane, if $f_t x \neq x$ for $t > 0$, or

(c) an immersed k-plane bundle over S^1, if x is on a periodic orbit.

(1.9) DEFINITION. A diffeomorphism with hyperbolic chain recurrent set is said to satisfy the *transversality condition* provided, for every $x, y \in R$, $W^s(x)$ intersects $W^u(y)$ transversally. A flow with hyperbolic R satisfies the transversality condition if, for every $x, y \in R$, $w^s(x)$ intersects $w^u(y)$ transversally.

Necessary and sufficient conditions for structural stability were conjectured by Smale [S1], and the sufficiency of these conditions were proved by J. Robbin [Ro] and C. Robinson [R1, R2].

(1.10) THEOREM. *A diffeomorphism or flow which has a hyperbolic chain recurrent set and satisfies the transversality condition is structurally stable.*

Whether or not the converse of this result is valid in general is an open question; however, Peixoto [P] proved quite early that both the theorem above and its converse hold for diffeomorphisms of the circle S^1 and flows on compact orientable surfaces. More recently Mañé [Ma] has proved the following

(1.11) THEOREM. *If $f: M^2 \to M^2$ is a structurally stable diffeomorphism of a compact two-dimensional manifold then f has a hyperbolic chain recurrent set R and satisfies the transversality condition.*

It is the hyperbolicity of R which is difficult to obtain as a consequence of structural stability. In fact it is easy to show that a structurally stable system with hyperbolic chain recurrent set must satisfy the transversality condition.

It is the results above which primarily motivate the hyperbolicity hypothesis; however hyperbolicity of R, even without the transversality condition, implies several other strong stability properties of the system. One of the most important of these is due independently to R. Bowen and C. Conley generalizing a result of D. V. Anosov.

(1.12) SHADOWING LEMMA [B2]. *Suppose $f: M \to M$ is a diffeomorphism with hyperbolic chain recurrent set R. Given $\epsilon > 0$ there exists a $\delta > 0$ such that, if $\{x_i\}_{i=-\infty}^{\infty}$ is any sequence in R satisfying $d(fx_i, x_{i+1}) < \delta$, then there is a unique $x \in R$ satisfying $d(x_i, f^i x) < \epsilon$ for all i.*

In other words if $\{x_i\}$ is a sufficiently good approximate orbit of f in R there is a unique true orbit of f which uniformly "ϵ-shadows" the approximate orbit $\{x_i\}$. Notice that if the sequence $\{x_i\}$ is periodic of period n ($x_i = x_{i+n}$ for all i), then the orbit of x is periodic of period n since the orbit of $y = f^n x$ also shadows $\{x_i\}$, and the uniqueness then implies $x = y$. This is one way to prove that the periodic points of f, Per(f), are dense in R, since by definition points of R lie on approximate periodic orbits (the proof requires the nontrivial fact that $R(f) = R(f|R)$ which can be found in [C]). A result for flows which is analogous to (1.12) is also valid.

A theorem of Smale [S2] says that if f is a diffeomorphism with $R(f)$ hyperbolic and g is any sufficiently small C^1 perturbation of f then f restricted to $R(f)$ is topologically conjugate to g restricted to $R(g)$ and $R(g)$ is hyperbolic with respect to g. The analogue of this theorem for flows with hyperbolic R was proved by Pugh and Shub [P-S].

The hyperbolicity of R also implies a result which gives a good description of how points in M approach R as time tends to ∞ or $-\infty$. In fact any such point is asymptotic not only to R, but to the orbit of a particular point of R. Hence for systems with hyperbolic R, if we understand the behavior of the system restricted to R then we understand the long run or asymptotic behavior of all points of M as well.

The following result was first proved in [HPPS]; a simpler proof was given by Bowen in (3.10) of [B1].

(1.13) THEOREM. *If a diffeomorphism or flow on M has hyperbolic chain recurrent set R then*

$$M = \bigcup_{x \in R} W^s(x) = \bigcup_{x \in R} W^u(x).$$

For systems with hyperbolic R we can say more about the Lyapunov functions mentioned in (1.2) and (1.3). From (1.2) it is clear that a Lyapunov function $g: M \rightarrow R$ must be constant on each basic set Λ_i since there is a dense orbit in Λ_i. Also from (2) of (1.2) it follows that $g(\Lambda_i) \neq g(\Lambda_j)$ if $\Lambda_i \neq \Lambda_j$. There is however a great deal of freedom in arranging the values of $g(\Lambda_i)$. One obvious restriction is that if

$$W^u(\Lambda_i) = \bigcup_{x \in \Lambda_i} W^u(x), \qquad W^s(\Lambda_j) = \bigcup_{x \in \Lambda_j} W^s(x),$$

and

$$y \in W^u(\Lambda_i) \cap W^s(\Lambda_j)$$

then $y \notin R$ so

$$g(\Lambda_j) = \lim_{t \to \infty} g(f_t(y)) < \lim_{t \to -\infty} g(f_t(y)) = g(\Lambda_i).$$

Hence, whenever $W^u(\Lambda_i) \cap W^s(\Lambda_j) \neq \varnothing$ then $g(\Lambda_i) > g(\Lambda_j)$. From the proof of existence of Lyapunov functions in [C] it is not difficult to see that this is the only restriction on the values of $g(\Lambda_i)$. We summarize this discussion in the following result.

(1.14) PROPOSITION. *If $g: M \rightarrow R$ is a Lyapunov function for a system with hyperbolic chain recurrent set R, then g is constant on each basic set Λ_i but separates basic sets. The function g can be chosen so the values $\{g(\Lambda_i)\}$, and in particular their order, are arbitrary except for the restriction*:

$$W^u(\Lambda_i) \cap W^s(\Lambda_j) \neq \varnothing \quad implies \quad g(\Lambda_i) > g(\Lambda_j).$$

(1.15) *Exercise.* Show that if $f: M \rightarrow M$ has a hyperbolic chain recurrent set and Λ is a basic set of index 0 or $n = \dim M$, then Λ consists of a single periodic orbit. Show the same for flows if the basic set Λ has index 0 or $n - 1$ and is not a rest point.

The last five lines were omitted from the first printing. Please substitute this page for the original page 7.

Chapter 2. Morse gradients

The simplest examples of dynamical systems with hyperbolic chain recurrent sets are the gradients of Morse functions. These are also the systems where the relationship between the dynamics of the system and homological invariants of the manifold on which it occurs are most transparent.

(2.1) DEFINITION. A C^∞ function $g: M \longrightarrow R$ is called a *Morse function* provided that each of its critical points is nondegenerate. A critical point $p \in M$ is *nondegenerate* if there exist local coordinates $(x_1, x_2, \ldots, x_k, y_1, \ldots, y_{n-k})$ centered at p such that $g(\vec{x}, \vec{y}) = g(p) - |\vec{x}|^2 + |\vec{y}|^2$ in these coordinates. The integer k is called the *index* of p. If M has boundary we assume that g is constant on each boundary component.

A flow f_t on M is called *gradient-like* with respect to g if: (1) The derivative $(d/dt)g(f_t x) < 0$ for all x which are not critical points of g. (2) On a neighborhood of the critical point p of g the vector field generating f_t is

$$-\nabla g = 2\left(\sum x_i \frac{\partial}{\partial x_i} - \sum y_j \frac{\partial}{\partial y_j}\right).$$

A theorem of Smale [S3] says that a gradient-like flow is, in fact, minus the gradient (with respect to some Riemannian metric) of a Morse function.

(2.2) PROPOSITION. *If f_t is gradient-like with respect to g then f_t has hyperbolic chain recurrent set and the basic sets are the critical points of g.*

The proof of this result is straightforward. Since $g(f_t x)$ is strictly decreasing if x is not a critical point, it follows that R is precisely the critical points of g, all of which are stationary points of f_t. Hyperbolicity at a critical point p is checked using the local coordinates $(\vec{x}, \vec{y})$ and the explicit formula in these coordinates

$$f_t(\vec{x}, \vec{y}) = (e^{2t}\vec{x}, e^{-2t}\vec{y})$$

which is valid as long as either side is contained in the domain U of the coordinates. Notice that $W^u(p) \cap U$ consists of the k dimensional manifold given by $\vec{y} = 0$, i.e., the points $\{(x_1, \ldots, x_k, 0, \ldots, 0) \in U\}$. Likewise, $W^s(p) \cap U$ is given by $\vec{x} = 0$.

The main result of this chapter is the following theorem obtained from results of M. Morse and S. Smale. For many manifolds it gives necessary and sufficient homological

Chapter 2. Morse gradients

The simplest examples of dynamical systems with hyperbolic chain recurrent sets are the gradients of Morse functions. These are also the systems where the relationship between the dynamics of the system and homological invariants of the manifold on which it occurs are most transparent.

(2.1) DEFINITION. A C^∞ function $g: M \to R$ is called a *Morse function* provided that each of its critical points is nondegenerate. A critical point $p \in M$ is *nondegenerate* if there exist local coordinates $(x_1, x_2, \ldots, x_k, y_1, \ldots, y_{n-k})$ centered at p such that $g(\vec{x}, \vec{y}) = g(p) - |\vec{x}|^2 + |\vec{y}|^2$ in these coordinates. The integer k is called the *index* of p. If M has boundary we assume that g is constant on each boundary component.

A flow f_t on M is called *gradient-like* with respect to g if: (1) The derivative $(d/dt)g(f_t x) < 0$ for all x which are not critical points of g. (2) On a neighborhood of the critical point p of g the vector field generating f_t is

$$-\nabla g = 2\left(\sum x_i \frac{\partial}{\partial x_i} - \sum y_j \frac{\partial}{\partial y_j} \right).$$

A theorem of Smale [S3] says that a gradient-like flow is, in fact, minus the gradient (with respect to some Riemannian metric) of a Morse function.

(2.2) PROPOSITION. *If f_t is gradient-like with respect to g then f_t has hyperbolic chain recurrent set and the basic sets are the critical points of g.*

The proof of this result is straightforward. Since $g(f_t x)$ is strictly decreasing if x is not a critical point, it follows that R is precisely the critical points of g, all of which are stationary points of f_t. Hyperbolicity at a critical point p is checked using the local coordinates $(\vec{x}, \vec{y})$ and the explicit formula in these coordinates

$$f_t(\vec{x}, \vec{y}) = (e^{2t}\vec{x}, e^{-2t}\vec{y})$$

conditions for the existence of a Morse function or gradient-like flow with a prescribed number of critical points of each index.

(2.3) THEOREM. *Suppose f_t is a gradient-like flow on a boundaryless manifold M with c_k rest points of index k. Then for every $k \geqslant 0$*

$$c_k - c_{k-1} + \cdots \pm c_0 \geqslant \beta_k(F) - \beta_{k-1}(F) + \cdots \pm \beta_0(F)$$

where $\beta_k(F) = \dim H_k(M; F)$, F a field. Conversely suppose $\{c_k\}$ is a set of nonnegative integers satisfying these inequalities for every field F. Then if M is simply connected and of dimension > 5, there exists a gradient-like flow on M with c_k critical points of index k.

The second half of this theorem is also valid for oriented surfaces. Its validity for simply connected three or four dimensional manifolds would imply the Poincaré conjecture in these dimensions.

This theorem is a paradigm for many of the results discussed in these lectures. In several settings we will first seek homological conditions necessary for the existence of a certain kind of dynamics on a given manifold or in a given homotopy class. Often these conditions are formally quite similar to the Morse inequalities of (2.3). Then we try to show that these conditions are sufficient, i.e., that any dynamical picture consistent with them can be realized.

The discussion leading to a proof of (2.3) actually gives us much more insight into the relationship between gradient-like flows and the homology of M than is apparent from the theorem. We proceed with the investigation of this relationship. First note that by slightly altering g we can assume that g is a Lyapunov function for f_t, specifically that for any two critical points p, q, with $p \neq q$, we have $g(p) \neq g(q)$. This alteration is achieved by choosing a smooth real valued function $\alpha(x)$ supported on a small neighborhood of p and having the value 1 on a smaller neighborhood of p. Then it is easily seen that if ϵ is sufficiently small, $g(x) + \epsilon\alpha(x)$ is a Morse function with the same critical points as g, f_t is gradient-like with respect to it. Hence by judicious choice of ϵ we can assure $g(p) + \epsilon\alpha(p) \neq g(q)$ for any critical point q.

The next step is to investigate how the sets $M(a) = g^{-1}((-\infty, a])$ change as we vary the real number a. It is a consequence of the implicit function theorem that $M(a)$ is a manifold provided a is not a *critical value,* i.e. the image under g of a critical point. A number b which is not a critical value is a *regular value.* One component of the boundary of $M(a)$ is $g^{-1}(a)$ unless $g^{-1}(a)$ is empty.

(2.4) LEMMA. *If $a < b$ are regular values and there are no critical points in $M(b) - M(a)$, then $M(a)$ is diffeomorphic to $M(b)$. Moreover, the manifold $g^{-1}([a, b])$ is diffeomorphic to $g^{-1}(b) \times [0, 1]$ or $g^{-1}(a) \times [0, 1]$.*

PROOF. Let $M(b, a) = g^{-1}([a, b])$. Since there is a negative upper bound for $(d/dt)g(f_t)|_{t=0}$ on $M(b, a)$, it follows that for each $x \in g^{-1}(b)$ there is a $\tau(x) > 0$ such that $f_{\tau(x)}(x) \in g^{-1}(a)$. The function $\tau(x)$ is smooth in x and hence the map $h: g^{-1}(b) \times [0, 1]$ $\rightarrow M(b, a)$ given by $h(x, s) = f_{s\tau(x)}(x)$ is a diffeomorphism (its inverse is given by

$h^{-1}(y) = (f_{\alpha(y)}(y), -\alpha(y))$, where for each $y \in M(b, a)$, $\alpha(y) < 0$ denotes the unique number such that $f_{\alpha(y)}(y) \in g^{-1}(b))$. The proof that $M(b, a) \cong g^{-1}(a) \times [0, 1]$ is similar.

It follows that $M(b, a - \epsilon) \cong M(a - \epsilon) \times [0, 1] \cong M(a, a - \epsilon)$ for sufficiently small $\epsilon > 0$, and with a small amount of care one can arrange that the diffeomorphism from $M(a, a - \epsilon)$ to $M(b, a - \epsilon)$ is the identity on a neighborhood of $g^{-1}(a - \epsilon)$. Thus it extends to a diffeomorphism from $M(a)$ to $M(b)$.   Q.E.D.

Evidently changes in $M(a)$ occur only when a passes through a critical value. To analyze this change we return to the local coordinates $(\vec{x}, \vec{y})$ defined on a neighborhood U of the critical point p. Since we are assuming that g separates all critical points we can choose U so that $g(U)$ is disjoint from $g(q)$ for all critical points $q \neq p$. If $c = g(p)$ we want to investigate the difference in the topology of $M(c - \epsilon)$ and $M(c + \epsilon)$ for small $\epsilon > 0$.

(2.5) DEFINITION. A *handle* for the critical point p of the Morse function g or of the flow f_t is a set $H(\epsilon)$ of the form

$$H(\epsilon) = \{(\vec{x}, \vec{y}) \in U \mid |\vec{y}|^2 \leqslant \epsilon, |\vec{x}|^2 \leqslant 2\epsilon\}.$$

It is a neighborhood of p diffeomorphic to $D^k \times D^{n-k}$ where k is the index of p and $n = \dim M$.

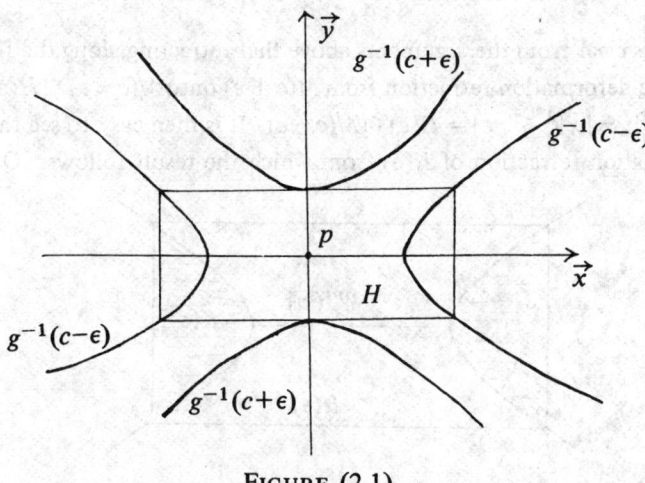

FIGURE (2.1)

(2.6) PROPOSITION. *The Manifold $M(c + \epsilon)$ is homeomorphic to $M(c - \epsilon) \cup H(\epsilon)$.*

PROOF.   The proof is essentially the same as the proof of (2.4). Let $M'(\epsilon, r), 0 < r \leqslant 1$, denote $M(c - \epsilon) \cup \{(\vec{x}, \vec{y}) \mid |\vec{x}|^2 \leqslant (1 + r)\epsilon$ and $|\vec{y}|^2 \leqslant r\epsilon\}$, so $M'(\epsilon, 1) = M(c - \epsilon) \cup H(\epsilon)$. The part of the boundary $\partial M'(\epsilon, r)$ which is in U is

$$\{(\vec{x}, \vec{y}) \mid |\vec{y}|^2 - |\vec{x}|^2 = -\epsilon \text{ and } |\vec{y}|^2 \geqslant r\epsilon\} \cup \{(\vec{x}, \vec{y}) \mid |\vec{y}|^2 = r\epsilon, |\vec{x}|^2 \leqslant (1 + r)\epsilon\}.$$

If we let $X(\epsilon, r)$ denote the component of $\partial M'(\epsilon, r)$ which intersects U, then using the flow lines of f_t, we can, as we did in (2.4), construct homeomorphisms

$$h_1: (M(c + \epsilon) - \text{int } M'(2\epsilon, 1/4)) \longrightarrow X(2\epsilon, 1/4) \times [0, 1],$$

$$h_2: (M'(\epsilon, 1) - \text{int } M'(2\epsilon, 1/4)) \longrightarrow X(2\epsilon, 1/4) \times [0, 1]$$

which are the identity on $X(2\epsilon, 1/4)$. Hence $h_2^{-1} \circ h_1$ extends by the identity to a homeomorphism from $M(c + \epsilon)$ to $M'(\epsilon, 1) = M(c - \epsilon) \cup H(\epsilon)$.   Q.E.D.

Two remarks concerning this result and its proof are in order. First, if one is willing to make the effort to "round the corners" on $M(c - \epsilon) \cup H(\epsilon)$ then it can be shown to be diffeomorphic to $M(c + \epsilon)$. Secondly $H(\epsilon)$ as defined above intersects the interior of $M(c - \epsilon)$. However, if we define $H'(\epsilon)$ to be $\{(\vec{x}, \vec{y}) \mid |\vec{y}|^2 \leqslant \epsilon$ and $|\vec{x}|^2 \leqslant |\vec{y}|^2 + \epsilon\}$ then it is easy to see that $H'(\epsilon)$ is diffeomorphic to $H(\epsilon)$ and $M(c - \epsilon) \cup H(\epsilon) = M(c - \epsilon) \cup H'(\epsilon)$, but $H'(\epsilon) \cap M(c - \epsilon)$ is a copy of $S^{k-1} \times D^{n-k}$ embedded in $\partial M(c - \epsilon)$. Thus one often says that $M(c + \epsilon)$ is obtained by attaching a k-handle $H'(\epsilon) \cong D^k \times D^{n-k}$ to $M(c - \epsilon)$ by an embedding of $\partial D^k \times D^{n-k}$ in $\partial M(c - \epsilon)$.

If we let $W_\epsilon^u(p)$ denote $\{(\vec{x}, \vec{y}) \mid \vec{y} = 0, |\vec{x}|^2 \leqslant \epsilon\}$ then $W_\epsilon^u(p)$ is a k dimensional disk neighborhood of p in $W^u(p)$ and $W_\epsilon^u(p) \cap M(c - \epsilon)$ is a $(k - 1)$ dimensional sphere embedded in $\partial M(c - \epsilon)$.

(2.7) COROLLARY. *The manifold $M(c + \epsilon)$ has $M(c - \epsilon) \cup W_\epsilon^u(p)$ as a strong deformation retract.*

PROOF. It is clear from the arguments above that retracting along the flow lines of f_t one can obtain a strong deformation retraction from $M(c + \epsilon)$ onto $M(c - \epsilon) \cup H(\epsilon)$. Let $K(\epsilon) = \{(\vec{x}, \vec{y}) \in H(\epsilon) \mid |\vec{y}|^2 - |\vec{x}|^2 \leqslant -\epsilon\} = H(\epsilon) \cap M(c - \epsilon)$. It is then easy to see that $K(\epsilon) \cup W_\epsilon^u(p)$ is a strong deformation retraction of $H(\epsilon)$ from which the result follows.   Q.E.D.

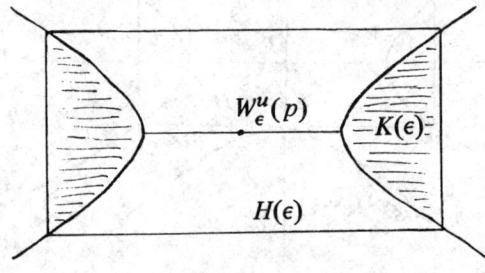

FIGURE (2.2)

We now consider gradient-like flows which satisfy the transversality condition (1.9). The following theorem of Smale [S3] is basic.

(2.8) THEOREM [S3]. *In the space of gradient-like flows with the C^r topology on a manifold M, those which satisfy the transversality condition form a dense open subset. If f_t is in this subset then it is gradient-like with respect to a Morse function g with the property that $g(p) = \text{index}(p)$ for every critical point p.*

A Morse function g with the property that $g(p) = \text{index}(p)$ for every critical point p is called *self-indexing*. It is interesting to compare the existence of such a g with the restrictions on Lyapunov functions from (1.14) and note that the transversality condition implies $W^s(p) \cap W^u(q) = \emptyset$ for dimensional reasons whenever index(p) $\leqslant$ index(q).

We can again define $M(c) = g^{-1}((-\infty, c])$, where g is now a self-indexing Morse function with respect to which f_t is gradient-like and we then have analogues of (2.6) and (2.7).

(2.9) PROPOSITION. *Suppose f_t is a gradient-like flow with respect to the self-indexing Morse function g and $\{p_1, \ldots, p_m\}$ are the critical points of index k. Then $M(k + \epsilon)$ is homeomorphic to $M(k - \epsilon) \cup (\bigcup_{i=1}^m h_i(\epsilon))$, where $h_i(\epsilon)$ is a handle for the critical point p_i, and $M(k + \epsilon)$ has $M(k - \epsilon) \cup (\bigcup_{i=1}^m W_\epsilon^u(p_i))$ as a strong deformation retract.*

The proof is essentially the same as the proofs of (2.6) and (2.7).

The picture of gradient-like flows which emerges from this discussion is one we will seek to recreate with much more complicated dynamical systems. The basic sets of Morse gradients are extremely simple—each consisting of a single rest point, and the unstable manifolds (or stable manifolds) of these points form a decomposition of the manifold into open cells $W^u(p)$ whose dimension is the index of p. This is reflected homologically by the following two results concerning the filtration defined by

$$M_k = g^{-1}((-\infty, k + 1 - \epsilon]), \qquad 0 \leqslant k \leqslant n = \dim M,$$

where g is a self-indexing Morse function.

(2.10) PROPOSITION. *If $\{p_1, \ldots, p_m\}$ are the critical points of index k, then*

$$H_i(M_k, M_{k-1}) \cong \begin{cases} Z^m & \text{if } i = k, \\ 0 & \text{otherwise.} \end{cases}$$

Moreover, if we denote by $[W^u(p_j)]$ the image of a generator of $H_k(W^u(p_j), W^u(p_j) \cap M_{k-1})$ in $H_k(M_k, M_{k-1})$ then $\{[W^u(p_j)]\}$ form a basis of $H_k(M_k, M_{k-1})$.

PROOF. By (2.4) and (2.9), it is clear there is a strong deformation retraction from M_k to $M(k - \epsilon) \cup (\bigcup_{j=1}^m W_\epsilon^u(p_j))$. It follows that

$$H_i(M_k, M_{k-1}) \cong H_i\left(M(k-\epsilon) \cup \left(\bigcup W_\epsilon^u(p_j)\right), M(k-\epsilon)\right)$$

$$\cong \bigoplus_j H_i(W_\epsilon^u(p_j), \partial W_\epsilon^u(p_j)) \quad \text{by excision}$$

$$\cong Z^m.$$

Since the retraction fixed $W^u(p_j)$ it is easy to check $[W^u(p_j)]$ form a basis.   Q.E.D.

(2.11) PROPOSITION. *If $C_k = H_k(M_k, M_{k-1}; Z)$ and $d_k: C_k \rightarrow C_{k-1}$ is the boundary map $\partial_k: H_k(M_k, M_{k-1}) \rightarrow H_{k-1}(M_{k-1}, M_{k-2})$ for the triple (M_k, M_{k-1}, M_{k-2}) then $\{C_k, d_k\}$ is a free chain complex whose homology is $H_*(M)$.*

This is a general fact about filtrations for which $H_i(M_k, M_{k-1}) = 0$ if $i \neq k$. An elementary proof can be found in (7.2) of [M1] or it can be proved trivially using the spectral sequence associated to the filtration.

The following classical result of M. Morse provides half of the main result of this chapter. We formulate it in terms of gradient-like flows.

(2.12) THEOREM (MORSE INEQUALITIES). *Suppose the flow f_t is gradient-like with respect to a Morse function g on the compact manifold M and V is the union of those components of ∂M on which the flow is exiting (V may be empty). If there are c_k critical points of index k and $\beta_k = \dim H_k(M, V; F)$ for a field F, then*

$$c_k - c_{k-1} + \cdots \pm c_0 \geqslant \beta_k - \beta_{k-1} + \cdots \pm \beta_0 \quad \text{for all } k \geqslant 0.$$

As a consequence $\Sigma(-1)^i c_i = \chi(M, V)$, the Euler characteristic.

PROOF. We first note that (2.8), the result of Smale which says gradient-like flows satisfying the transversality condition are dense and open in the set of all gradient-like flows, allows us to perturb f_t to satisfy this condition and assume g is self-indexing. (A small C^1 perturbation will not change the number of critical points of each index.) The result then follows immediately from (2.10), (2.11) and the following proposition.

(2.13) PROPOSITION. *Given a set of nonnegative integers $\{\beta_0, \beta_1, \ldots, \beta_n\}$, then necessary and sufficient conditions on a set of nonnegative integers $\{c_i\}$ for the existence of a finite chain complex C over the field F with $\dim C_i = c_i$ and $\dim H_i(C) = \beta_i$ are that for all $k \geqslant 0$*

$$c_k - c_{k-1} + \cdots \pm c_0 \geqslant \beta_k - \beta_{k-1} + \cdots \pm \beta_0.$$

PROOF. We first prove sufficiency. Define a complex $D(k)$ by

$$D_i(k) = 0 \quad \text{if } i \neq k, k-1,$$

$$D_k(k) = Z_k \oplus B_k, \qquad D_{k-1}(k) = B_k$$

where $\dim Z_k = \beta_k$, $\dim B_k = (c_k - \cdots \pm c_0) - (\beta_k - \cdots \pm \beta_0)$ and $\partial_k : D_k(k) \longrightarrow D_{k-1}(k)$ is given by $\partial_k(Z_k) = 0$, $\partial_k : B_k \longrightarrow B_k$ is the identity. Now let $C = \bigoplus_k D(k)$. Then $H_*(C) = \bigoplus_k H_*(D(k))$ so $H_j(C) \cong Z_j$ which has dimension β_j as desired. Also, $\dim C_j = \beta_j + \dim B_j + \dim B_{j-1} = c_j$ as desired.

To prove necessity we use the following algebraic lemma which can be found in [L, p. 99]. It and its generalizations will be used extensively in these lectures.

(2.14) LEMMA. *If $\{C_i, \partial_i\}_{i=0}^n$ is a chain complex of finite dimensional vector spaces then*

$$\sum_{i=1}^n (-1)^i \dim C_i = \sum_{i=1}^n (-1)^i \dim H_i(C).$$

Consider the complex given by

$$0 \longrightarrow \partial_{k+1}(C_{k+1}) \overset{i}{\longrightarrow} C_k \overset{\partial_k}{\longrightarrow} C_{k-1} \longrightarrow \cdots \longrightarrow C_0 \longrightarrow 0,$$

where i is inclusion. Applying (2.14) to this we get

$$\sum_{j=0}^{k} (-1)^j \dim C_j + (-1)^{k+1} \dim \partial(C_{k+1}) = \sum_{j=0}^{k} (-1)^j \beta_j.$$

Since $\dim \partial(C_{k+1}) \geq 0$ this gives $(-1)^k \Sigma_{j=0}^k (-1)^j \dim C_j \geq (-1)^k \Sigma_{j=0}^k (-1)^j \beta_j$ which is the desired inequality. Q.E.D.

We turn now to the converse problem of realizing gradient-like flows with a pre-assigned number of critical points of each index. The key tool for this purpose is the h-cobordism theorem of Smale [S4] (see also the excellent book [M1] devoted to the proof of this theorem).

(2.15) h-COBORDISM THEOREM [S4]. *Suppose $g_0: M \longrightarrow R$ is a Morse function with regular values $a < b$ and $M(b, a) = g_0^{-1}([a, b])$, $V(a) = g_0^{-1}(a)$, $V(b) = g_0^{-1}(b)$. Then if the dimension of M is > 5, $\Pi_1(M(b, a)) = \Pi_1(V(a)) = \Pi_1(V(b)) = 0$ and $H_*(M(b, a), V(a)) = 0$, there is a Morse function $g_1: M \longrightarrow R$ which agrees with g_0 except on the interior of $M(b, a)$ and which has no critical points in $M(b, a)$.*

Of course it then follows from (2.4) that $M(b, a)$ is diffeomorphic to $V(a) \times I$.

The following result though not explicitly stated is essentially contained in [S5]. We give a brief sketch of a somewhat different proof.

(2.16) THEOREM. *Suppose M is a compact simply connected manifold of dimension > 5, and C is a finitely generated free chain complex with $H_*(C) \cong H_*(M)$. Then there exists a self-indexing Morse function $g: M \longrightarrow R$ such that if $M_k = g^{-1}((-\infty, k + \frac{1}{2}])$ then the chain complex $\{H_k(M_k, M_{k-1}), \partial_k\}$ is isomorphic to C.*

Recall that $\partial_k: H_k(M_k, M_{k-1}) \longrightarrow H_{k-1}(M_{k-1}, M_{k-2})$ is the boundary map of the triple (M_k, M_{k-1}, M_{k-2}) and according to (2.11) $\{H_k(M_k, M_{k-1}), \partial_k\}$ is a chain complex whose homology is $H_*(M)$. The theorem at hand then says any chain complex C with $H_*(C) \cong H_*(M)$ can be realized in this way provided $\Pi_1(M) = 0$ and the dimension of $M > 5$.

The idea of the proof is to let D be some realizable chain complex, e.g., the one from (2.11), and use the fact that $H_*(D) \cong H_*(C)$ implies that $D \oplus E^1 \cong C \oplus E^2$ where E^1 and E^2 are the sum of acyclic elementary complexes (see (II, 4.3) of [D] and Theorem 9, p. 191 of [Wh]). Recall that a chain complex E is elementary provided, for some k, $E_k \cong E_{k+1} \cong Z$ and $E_i = 0$, $i \neq k$, $k + 1$, and it is in addition acyclic if $\partial_{k+1}: E_{k+1} \longrightarrow E_k$ is an isomorphism. Since D is realizable it is not difficult to realize $D \oplus E^1$ by successively altering g to add a pair of "canceling" critical points of index k and $k + 1$, and thereby adding an acyclic elementary complex with $E_k \cong E_{k+1} \cong Z$ to the chain complex corresponding to g. To do this we use the following.

(2.17) LEMMA. *Let $f: W \longrightarrow [a, b]$ be a Morse function defined on a connected n-dimensional manifold such that all critical points of index 0 are in $f^{-1}(a)$ and all critical points of index n are in $f^{-1}(b)$. Then given c, d such that $a < c < d < b$, and $0 \leq k < n$, there exists a Morse function $g: W \longrightarrow [a, b]$ such that*

(a) *The functions f and g agree on a neighborhood of ∂W and on a neighborhood of the critical points of f.*

(b) *The function g has two additional critical points: one of index k in $g^{-1}(c)$ and one of index $k + 1$ in $g^{-1}(d)$.*

(c) *The unstable manifolds of the additional critical points are disjoint from the stable manifolds of any of the previous critical points except perhaps those of index 0 and this is also true for the dual Morse function $-g$.*

PROOF. Because all critical points of f of index 0 and n are in $f^{-1}(a) \cup f^{-1}(b)$ and W is connected there is an integral curve I of $-\nabla f$ running from a point in $f^{-1}(d + \epsilon)$ to a point in $f^{-1}(c - \epsilon)$, for some small $\epsilon > 0$. We may assume I is disjoint from stable and unstable manifolds of all critical points of index $\neq 0, n$, since these manifolds will have dimension $< n$. By standard arguments there is a neighborhood U of I (also disjoint from stable and unstable manifolds) and a chart map $\phi: U \longrightarrow \phi(U) \subset R^n$ such that if $x \in U$, $f(x) = h(\phi(x))$ where $h: R^n \longrightarrow R$ is a linear height function, say $h(x_1, \ldots, x_n) = x_1$. By a linear change of coordinates in R we can assume $c = -d$, $d > 0$. We now alter h to $\hat{h}$ so that in a small neighborhood of the interval joining $(a, 0, \ldots, 0)$ and $(-a, 0, \ldots, 0)$ in R^n, $\hat{h}(x_1, \ldots, x_n) = (x_1^3 - 3a^2 x_1) - x_2^2 - \cdots - x_{k+1}^2 + x_{k+2}^2 + \cdots + x_n^2$, where a is the cube root of $d/2$. However we keep $\hat{h}(x_1, \ldots, x_n) = x_1$ outside of $\phi(U)$, and arrange that $\hat{h}$ has only the critical points at $\pm(a, 0, \ldots, 0)$ which are easily seen to have index k and $k + 1$ and critical values d and $c = -d$. More details of this construction can be found in (8.2) of [M1]. Finally $g: M \longrightarrow R$ is defined by

$$g(x) = \begin{cases} f(x) & \text{if } x \notin U, \\ \hat{h}(\phi(x)) & \text{if } x \in U. \end{cases} \qquad \text{Q.E.D.}$$

We return now to the sketch of the proof of (2.16).

Having realized $D \oplus E^1 \cong C \oplus E^2$ we note that using the h-cobordism theorem it is possible to alter g canceling critical points p and q of index k and $k + 1$ and thereby eliminate an acyclic elementary summand E of E^2, provided $2 \leqslant k \leqslant n - 3$ and that $[W^u(p)]$ and $[W^u(q)]$ represent generators of E_k and E_{k+1}. However, with the restriction $2 \leqslant k \leqslant n - 2$ it is possible to alter g so that $\{[W^u(p_i)] \mid p_i$ is a critical point of index $k\}$ represents any desired basis of $C \oplus E^2$ (see [M1]). Thus we can eliminate all the acyclic summands in dimensions k and $k + 1$ when $2 \leqslant k$ and $k + 1 \leqslant n - 2$. This is enough to prove the theorem provided $C = \{C_k, \partial_k\}$ satisfies $C_1 = C_{n-1} = 0$, $C_0 = C_n = Z$ (using the h-cobordism theorem techniques it is possible to choose D so that $D_0 = D_n = Z$, $D_1 = D_{n-1} = 0$). For the general case we note that since $H_1(C) = H_{n-1}(C) = 0$, $C \cong C' \oplus E$ where C' satisfies $C'_1 = C'_{n-1} = 0$, $C'_0 \cong C'_n \cong Z$ and E is the sum of acyclic elementary complexes, so we first realize C' and then add canceling critical points as before to realize C. This completes the outline of the proof of (2.16). A generalization of this result to nonsimply connected manifolds has been proved by M. Maller [Ma1].

We can now finish the proof of our main theorem (2.3).

(2.18) COROLLARY. *Suppose M is simply connected, has dimension* > 5, *and* $\beta_k(F) = \dim H_k(M; F)$. *If the nonnegative integers* $\{c_i\}$ *satisfy*

$$c_k - c_{k-1} + \cdots \pm c_0 \geqslant \beta_k(F) - \beta_{k-1}(F) + \cdots \pm \beta_0(F)$$

for every field F, then there exists a Morse function g on M with c_k *critical points of index k for each k.*

PROOF. By (2.16) it suffices to produce a free chain complex C with rank $C_k = c_k$ and $H_*(C) \cong H_*(M)$. We consider a free chain complex D with $H_*(D) = H_*(M)$ which has each D_k of minimal rank, i.e. with rank $D_k = \text{rank } H_k(M) + \tau_k + \tau_{k+1}$ where τ_k is the smallest number of cyclic summands possible for a decomposition of the torsion subgroup of $H_k(M)$ into a direct sum of cyclic summands. Now D, like every free chain complex, is a direct sum of elementary complexes, i.e. complexes E with $E_k \cong E_{k+1} \cong Z$ for some k and $E_i = 0$ if $i \neq k, k+1$. Let $\beta_k(F) = \dim H_k(D \otimes F)$ and suppose

$$c_k - c_{k-1} + \cdots \pm c_0 \geqslant \beta_k(F) - \beta_{k-1}(F) + \cdots \pm \beta_0(F)$$

for all k and every field F. We will construct a chain complex C with $H_*(C) \cong H_*(D)$ and $c_k = \text{rank of } C_k$ by inducting on d, the number of elementary complexes in a direct sum decomposition of D into elementary complexes.

If $d = 0$, so $H_*(D) = 0$, let C be a direct sum of $\alpha_k = c_k - c_{k-1} + \cdots \pm c_0$ acyclic elementary complexes of dimension $(k, k-1)$. Then $H_*(C) = 0$ and it is easy to see that rank of $C_k = \alpha_k + \alpha_{k-1} = c_k$.

To prove the induction step suppose $D \cong D' \oplus E$ where E is an elementary complex with $E_i = 0$ if $i \neq j, j+1$. Let $c_i' = c_i$ if $i \neq j, j+1$ and $c_j' = c_j - 1$, $c_{j+1}' = c_{j+1} - 1$. If $\partial_{j+1}: E_{j+1} \cong Z \rightarrow E_j \cong Z$ is multiplication by m and $\beta_k'(F) = \dim H_k(D' \otimes F)$ then

$$\beta_j'(F) = \begin{cases} \beta_j(F) - 1 & \text{if the characteristic of } F \text{ divides } m, \\ \beta_j(F) & \text{otherwise,} \end{cases}$$

and $\beta_{j+1}'(F) = \beta_{j+1}(F)$ or $\beta_{j+1}(F) - 1$ in the same fashion. Since $\beta_i'(F) = \beta_i(F)$ if $i \neq j, j+1$ one checks that $c_k' - c_{k-1}' + \cdots \pm c_0' \geqslant \beta_k'(F) - \beta_{k-1}'(F) + \cdots \pm \beta_0'(F)$ for all k and every field F. It then follows by the induction hypothesis that there exists a complex C' with $H_*(C') = H_*(D')$ and rank $C_k' = c_k'$ for all k. If we let $C = C' \oplus E$ then clearly $H_*(C) = H_*(C') \oplus H_*(E) = H_*(D') \oplus H_*(E) = H_*(D)$ and rank of $C_k = c_k$ as desired. Q.E.D.

Chapter 3. Symbolic dynamics for basic sets

The single most striking qualitative feature of diffeomorphisms with hyperbolic chain recurrent set is the symbolic dynamic structure they possess.

In general terms there are two major themes in the theory of smooth dynamical systems—statistical or probabilistic methods and topological methods. While often these two approaches differ greatly in their applications and the kind of information they give us about a dynamical system, there is a strong unifying element between them in the form of symbolic dynamics. We will not deal with the ergodic theory of symbol shifts in these lectures, but it is central to our investigation to understand how symbol shifts arise in dynamical systems and how they are related to homological invariants of the system. This chapter is devoted to a description of how symbol shifts occur in systems with hyperbolic chain recurrent set and some of their properties.

We begin with an example of a diffeomorphism of the two-sphere S^2 which we think of as the plane with a point ∞ at infinity added. In the plane we choose a region X consisting of three disks and two strips and map it as shown in Figure (3.1). The map is defined so that the points $\{p_1, p_2, p_3\}$ form a periodic attractor of period 3. We also impose conditions on the behavior of the map on the two strips. Each strip is foliated in two ways—by horizontal and by vertical line segments.

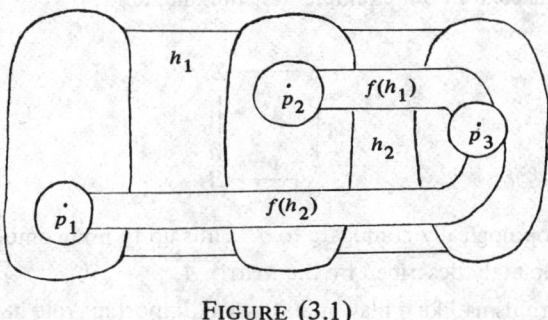

FIGURE (3.1)

We arrange that f uniformly stretches each horizontal line segment and that f(horizontal line segment) contains any horizontal line segment it intersects. Analogously we arrange that f uniformly contracts vertical line segments and that f(vertical line segment) is contained in any vertical line segment it intersects. Finally we extend the map to all of S^2, making ∞ an expanding fixed point, in such a way that the forward orbit of every point except ∞ enters X.

We now consider the asymptotic behavior of points under the diffeomorphism f. Many points x will satisfy $f^{-n}(x) \to \infty$ as $n \to \infty$ and many will satisfy $f^n(x) \to \{p_1, p_2, p_3\}$ as $n \to \infty$. We consider the closed invariant set Λ of points which do neither. These are the points whose entire orbit lies in the two strips h_1 and h_2. It is clear that the horizontal stretching and vertical contraction of f on $h_1 \cup h_2$ makes Λ a hyperbolic set with E^u the subbundle of TM_Λ of horizontal vectors and E^s the subbundle of vertical vectors. We will show below that Λ consists of chain recurrent points and is in fact a basic set of f. The other two basic sets of f are the single point ∞ and $\{p_1, p_2, p_3\}$.

It turns out that each point in Λ is determined by specifying which strip it is in after n iterates of f, $n \in Z$. However, it will be more convenient to specify a point by telling which strip it is in after n iterates of f^{-1}. Thus if we give $\{1, 2\}$ the discrete topology and consider the space of bi-infinite sequences $\Pi_{-\infty}^\infty \{1, 2\}$ with the product topology, there is a map $\Psi \colon \Lambda \to \Pi_{-\infty}^\infty \{1, 2\}$ defined by $\Psi(x) = \mathbf{a} = (\dots, a_{-1}, a_0, a_1, \dots)$ where for each $n \in Z$

$$
a_n = \begin{cases} 1 & \text{if } f^{-n}(x) \in h_1, \\ 2 & \text{if } f^{-n}(x) \in h_2. \end{cases}
$$

It is clear that the map Ψ is not surjective because $f(h_1) \cap h_1 = \varnothing$, so $a_n = 1$ implies $a_{n+1} = 2$. We will see, however, that this is the only restriction on the sequences in $\Psi(\Lambda)$ and we can codify it in a matrix. We let $A_{ij} = 1$ if $h_i \cap f(h_j) \neq \varnothing$ and 0 otherwise, so $A = \left(\begin{smallmatrix} 0 & 1 \\ 1 & 1 \end{smallmatrix} \right)$. If we define $\Sigma_A \subset \Pi_{-\infty}^\infty \{1, 2\}$ to be $\{\mathbf{a} \,|\, \text{if } a_i = 1 \text{ then } a_{i+1} = 2\}$, then Ψ is a homeomorphism onto Σ. A proof of this in greater generality will be given below in (3.11).

There is also a homeomorphism $\sigma \colon \Sigma \to \Sigma$, which shifts to the right, defined by $\sigma(\mathbf{a}) = \mathbf{b}$ where $b_n = a_{n-1}$. It is clear in our example that the diagram

$$
\begin{array}{ccc}
\Lambda & \xrightarrow{\Psi} & \Sigma \\
\downarrow{\scriptstyle f} & & \downarrow{\scriptstyle \sigma} \\
\Lambda & \xrightarrow{\Psi} & \Sigma
\end{array}
$$

commutes, so $f|\Lambda$ is topologically conjugate to σ. Thus up to homeomorphism the dynamics of orbits in Λ are completely described by the matrix A.

Shift homeomorphisms like σ play an extremely important role in the systems we are investigating, so at this point we give a precise definition and examine their properties. Suppose S is a finite set (whose elements we call "symbols") provided with the discrete topology and a relation $\to$ (which we think of as "can be followed by").

(3.1) DEFINITION. The *subshift of finite type* determined by the set S and relation $\to$ is the homeomorphism $\sigma \colon \Sigma \to \Sigma$, where $\Sigma \subset \Pi_{-\infty}^\infty S$ is defined by $\Sigma = \{\mathbf{s} = (\dots, s_{-1}, s_0, s_1, \dots) \,|\, s_i \to s_{i+1} \text{ for all } i\}$ and $\sigma(\mathbf{s}) = \mathbf{s}'$ where $s_i' = s_{i-1}$, so σ shifts to the right.

Subshifts of finite type are often specified by giving oriented graphs or by giving non-negative integer matrices. There are two somewhat different correspondences between graphs or matrices and subshifts of finite type. Both of them are of use to us.

Vertex graphs and 0-1 matrices. Given the set S and relation $\longrightarrow$ we form the oriented graph G with one vertex for each element s of S and an oriented edge from s to s' provided $s \longrightarrow s'$. Suppose each edge of G has length one and let Γ be the set of orientation and arc length preserving paths $\gamma \colon R \longrightarrow G$ such that $\gamma(n)$ is a vertex for each integer n. Then the correspondence $\gamma \longrightarrow (\dots, \gamma(-1), \gamma(0), \gamma(1), \dots) \in \Sigma$ defines a homeomorphism from Γ with the compact open topology to Σ. This homeomorphism gives a topological conjugacy between $\sigma \colon \Sigma \longrightarrow \Sigma$ and $\rho \colon \Gamma \longrightarrow \Gamma$ where $\rho(\gamma)(t) = \gamma(t - 1)$ for all t.

From the set S and relation $\longrightarrow$ (or from G) we can also construct a matrix of zeroes and ones. We number the elements of S, say from 1 to n, and let A be the $n \times n$ matrix given by

$$A_{ij} = \begin{cases} 1 & \text{if } s_i \longrightarrow s_j, \\ 0 & \text{otherwise.} \end{cases}$$

We call σ the *vertex shift* associated to the matrix A.

Of course given an $n \times n$ matrix of zeroes and ones (or an oriented graph with at most one edge running between each ordered pair of vertices) one can reconstruct the subshift of finite type, by letting $S = \{1, \dots, n\}$ and defining $i \longrightarrow j$ if and only if $A_{ij} = 1$. The subshift of finite type obtained from A is denoted $\sigma(A) \colon \Sigma_A \longrightarrow \Sigma_A$. In the example beginning this chapter we showed a basic set Λ with $f|\Lambda$ topologically conjugate to $\sigma(A)$: $\Sigma_A \longrightarrow \Sigma_A$ where $A = \left(\begin{smallmatrix} 0 & 1 \\ 1 & 1 \end{smallmatrix}\right)$.

A drawback of this scheme is that we often want to consider matrices whose entries are greater than one or graphs with several edges joining the same pair of vertices. This difficulty is rectified by a slightly different way of assigning a subshift to a matrix.

Edge graphs and matrices. Given an oriented graph G let S be the set of *edges* of G and define $s \longrightarrow s'$ provided s ends in the vertex from which s' emanates. In this way we associate a shift $\sigma \colon \Sigma \longrightarrow \Sigma$ to the graph G. We define $\rho \colon \Gamma \longrightarrow \Gamma$ exactly as before by $\rho(\gamma)(t) = \gamma(t - 1)$ for $\gamma \in \Gamma$, the space of orientation and arc length preserving paths in G with $\gamma(n)$ a vertex for each integer n. We can now define a topological conjugacy $h \colon \Gamma \longrightarrow \Sigma$ by

$$h(\gamma) = (\dots, s_{-1}, s_0, s_1, \dots) \in \Sigma \quad \text{where } s_i = \gamma([i, i + 1]).$$

It is clear that $h \circ \rho = \sigma \circ h$ and that h is a homeomorphism. When using the vertex graphs above we specified a path in G by giving the sequence of vertices it passes through; the approach we now take is to specify the path by giving a sequence of edges.

With the graph G we can also associate a nonnegative integer matrix A by numbering the vertices, say 1 to n, and defining an $n \times n$ matrix by A_{ij} = the number of edges going from vertex i to vertex j. Of course given the matrix A we can reconstruct the graph or we can go directly to the subshift as follows. We construct a finite set S and two functions l, r: $S \longrightarrow \{1, 2, \ldots, n\}$ by requiring $S = \bigcup S_{ij}$, the cardinality of $S_{ij} = A_{ij}$, and if $x \in S_{ij}$ then $l(s) = i$, $r(s) = j$. The relation $\longrightarrow$ on S is given by $s \longrightarrow s'$ if and only if $r(s) = l(s')$. In other words, given A we construct $\Sigma_{ij} A_{ij}$ symbols, with A_{ij} of them subscripted ij, and we say s' can follow s provided the right subscript of s is the same as the left subscript of s'. It is clear that this gives the same shift as the edge graph obtained by taking n vertices and joining vertex i to vertex j with A_{ij} edges. We will call this subshift of finite type the *edge shift* associated to A.

Given a matrix A of zeroes and ones the edge shift and the vertex shift associated to it are different in general (they may have different numbers of symbols for example). However, the following result permits us to avoid carefully distinguishing them in most cases.

(3.2) PROPOSITION. *If A is an $n \times n$ matrix of zeroes and ones then the edge shift $\sigma_e \colon \Sigma_e \longrightarrow \Sigma_e$ and the vertex shift $\sigma_v \colon \Sigma_v \longrightarrow \Sigma_v$ associated to A are topologically conjugate.*

PROOF. Let G be the graph with vertices labelled 1 to n and an oriented edge going from i to j if $A_{ij} = 1$. As above let Γ be the space of paths on G and define $\rho \colon \Gamma \longrightarrow \Gamma$ by $\rho(\gamma)(t) = \gamma(t-1)$. In the discussion above we showed that ρ is topologically conjugate to σ_e and to σ_v. It follows that σ_e is topologically conjugate to σ_v.   Q.E.D.

Henceforth we will generally refer to the shift associated to the matrix A without specifying whether we mean the edge shift or the vertex shift, since it will usually make no difference. Of course if A has entries > 1 only the edge shift makes sense.

(3.3) PROPOSITION. *If $\sigma(A)$ denotes the subshift corresponding to A, then for any $k > 0$, $\sigma(A^k)$ is topologically conjugate to $(\sigma(A))^k$, and $(\sigma(A))^{-1}$ is topologically conjugate to $\sigma(A^t)$ where A^t denotes the transpose of A.*

PROOF. We use the edge shifts corresponding to A and A^k. Let S be the symbols for A and r, $l \colon S \longrightarrow \{1, \ldots, n\}$ the "subscript functions", so $s \longrightarrow s'$ if and only if $r(s) = l(s')$. Similarly let P be the set of symbols for A^k and r', $l' \colon P \longrightarrow \{1, \ldots, n\}$ the "subscript functions". If $P_{ij} = \{p \in P | l'(p) = i, r'(p) = j\}$ then the cardinality of $P_{ij} = (A^k)_{ij}$. If $S_{ij}^k = \{(s_1, \ldots, s_k) \in S^k | l(s_1) = i, r(s_k) = j, r(s_t) = l(s_{t+1}), 1 \leqslant t < k\}$ (this is just the paths on G of length k starting at i and ending at j), it is easy to see that the cardinality of S_{ij}^k is also $(A^k)_{ij}$. Hence if $S(k) = \bigcup S_{ij}^k = \{(s_1, \ldots, s_k) \in S^k | l(s_t) = r(s_{t+1})$ for $1 \leqslant t < k\}$ then we can find a bijection $\Phi \colon P \longrightarrow S(k)$ such that $\Phi(P_{ij}) = S_{ij}^k$. If $(s_1, \ldots, s_k)$ and $(s_1', \ldots, s_k')$ are in $S(k)$ and we say $(s_1, \ldots, s_k) \longrightarrow (s_1', \ldots, s_k')$ if $r(s_k) = l(s_1')$ then it is clear that the set $S(k)$ with relation $\longrightarrow$ determines a shift topologically conjugate to $(\sigma(A))^k$. On the other hand the bijection $\Phi \colon P \longrightarrow S(k)$ respects the relations $\longrightarrow$ on P and $S(k)$ so it determines a topological conjugacy between the shift associated to A^k and the shift associated to $S(k)$ and $\longrightarrow$. Thus $(\sigma(A))^k$ is topologically conjugate to $\sigma(A^k)$.   Q.E.D.

(3.4) COROLLARY. *If $\sigma \colon \Sigma_A \longrightarrow \Sigma_A$ is the shift associated to A then the cardinality of the fixed point set $\mathrm{Fix}(\sigma^n)$ equals $\mathrm{tr}(A^n)$ for all $n > 0$.*

PROOF. If σ is the edge shift associated to A then $s = (\ldots, s_{-1}, s_0, s_1, \ldots)$ is a fixed point of σ if and only if $s_i = s_0$ for all i. Thus there is one such s for each $s_0 \in S$ satisfying $r(s_0) = l(s_0)$, and

$$\text{card Fix}(\sigma) = \sum_i \text{card } S_{ii} = \sum_i A_{ii} = \text{tr } A.$$

This proves the result for $n = 1$. The general result now follows since σ^n is topologically conjugate to $\sigma(A^n)$, $n > 0$, and $\text{Fix}(\sigma^{-n}) = \text{Fix}(\sigma^n)$.   Q.E.D.

In the example which began this chapter we claimed that the set Λ was a basic set. In order to prove this it is necessary to show that the shift conjugate to $f|\Lambda$ consists of chain recurrent points and has a dense orbit. Neither of these properties is true for an arbitrary shift.

(3.5) DEFINITION. A nonnegative $n \times n$ integer matrix is called *irreducible* provided for each $1 \leqslant i, j \leqslant n$ there is a $k > 0$ with $(A^k)_{ij} \neq 0$.

Suppose A is irreducible and K is the maximum of the finite set $\{k(i, j) | k(i, j)$ is the smallest k with $(A^k)_{ij} \neq 0\}$. Then any symbol can eventually follow any other, and in fact, given s, s' there exist $s_1, s_2, \ldots, s_l$, $l \leqslant K$, such that $s \to s_1 \to s_2 \to \cdots \to s_l \to s'$. Conversely if any symbol can eventually follow any other in the shift associated to the matrix A, then A is irreducible, indeed $(A^k)_{ij}$ is the number of admissible strings of symbols of length k starting with a symbol whose left subscript is i and ending with a symbol whose right subscript is j. The following result is an easy exercise based on this discussion.

(3.6) PROPOSITION. *If $\sigma: \Sigma \to \Sigma$ is the shift associated with the matrix A then the following are equivalent:*
   (a) *The matrix A is irreducible.*
   (b) *The homeomorphism σ has a dense orbit.*
If either of these hold the periodic points of σ are dense in Σ.

In the example with which this chapter began the matrix A of the shift $\sigma(A): \Sigma_A \to \Sigma_A$ is irreducible. Hence $f: \Lambda \to \Lambda$ has a dense orbit and the periodic points of f are dense in Λ.

If a matrix A is not irreducible we can still describe the chain recurrent set of $\sigma(A)$. For example if $A = \begin{pmatrix} B & W \\ 0 & C \end{pmatrix}$ where B and C are $m \times m$ and $n \times n$ irreducible matrices respectively, then

$$X = \{c \in \Sigma_A | 1 \leqslant r(c_i) \leqslant m, \ 1 \leqslant l(c_i) \leqslant m, \text{ for all } i\}$$

is a closed invariant set and $\sigma(A)|X$ is topologically conjugate to $\sigma(B)$. Likewise there is a closed invariant set Y with $\sigma(A)|Y$ conjugate to $\sigma(C)$. Consideration of the edge graph associated to A shows that W represents nonrecurrent orbits going from Y to X. The chain recurrent set of $\sigma(A)$ consists of $X \cup Y$. The following result shows that this behavior is typical.

(3.7) PROPOSITION. *If $\sigma\colon \Sigma \longrightarrow \Sigma$ is a subshift of finite type then it is topologically conjugate to the shift associated to a matrix A of the form*

$$\begin{pmatrix} A_1 & * & \cdot & \cdot & \cdot & * \\ 0 & A_2 & & & & \\ & & \cdot & & & \\ & & & \cdot & & * \\ & & & & \cdot & \\ 0 & & & 0 & & A_n \end{pmatrix}$$

where each A_i is irreducible. Thus the chain recurrent pieces are topologically conjugate to the shifts associated to the A_i's.

PROOF. Form an edge graph G associated to σ and define a relation $\leqslant$ on its vertices by saying $v \leqslant v'$ if there is an oriented path running from v' to v. We say that vertices v and v' are equivalent if $v \leqslant v'$ and $v' \leqslant v$ or if $v = v'$. Now number the vertices in such a way that vertices in the same equivalence class are numbered consecutively and if $v \leqslant v'$ but $v' \not\leqslant v$ then the number of v is less than the number of v'. Let A be the matrix given by A_{ij} = the number of edges running from vertex i to vertex j. Clearly σ is topologically conjugate to $\sigma(A)\colon \Sigma_A \longrightarrow \Sigma_A$. Also it is easy to see that A has the desired form. Each A_i corresponds to a subgraph of G consisting of the vertices in one equivalence class and the edges joining them. Q.E.D.

Quite different matrices can give rise to topologically conjugate subshifts of finite type. The equivalence relation on matrices which corresponds to topological conjugacy of the subshifts was discovered by Williams [Wm1].

(3.8) DEFINITION. Two nonnegative square integer matrices A and B are said to be *strong shift equivalent* provided there exist nonnegative integer matrices (not necessarily square) R_i and S_i, $1 \leqslant i \leqslant n$, such that $A = R_1 S_1$, $B = S_n R_n$ and $R_{i+1}S_{i+1} = S_i R_i$ for $1 \leqslant i < n$.

In other words, this is just the equivalence relation on matrices generated by the relation $RS \sim SR$. A proof of the following theorem of Williams is given in an appendix.

(3.9) THEOREM [Wm1]. *Suppose A and B are nonnegative square integer matrices and $\sigma(A)$, $\sigma(B)$ are the corresponding subshifts of finite type. Then $\sigma(A)$ is topologically conjugate to $\sigma(B)$ if and only if A is strong shift equivalent to B.*

We return now to the example which began this chapter and prove a result which will establish the claim that the basic set contained in the handles $h_1 \cup h_2$ is a subshift of finite type.

Suppose $f\colon M \longrightarrow M$ is a diffeomorphism of an n-dimensional manifold and $H = \bigcup h_i$ is a finite set of disjoint k-handles in M. That is, each h_i is an embedded copy of $D^k \times D^{n-k}$ (we will write $h_i = D_i^k \times D_i^{n-k}$). If x is a point of $D_i^k \times p \subset h_i$ we will denote the k-disk $D_i^k \times p$ in h_i by $W_i^u(x)$. Similarly $W_i^s(x)$ will denote the $(n-k)$-disk $q \times D_i^{n-j}$ containing x.

(3.10) DEFINITION. The diffeomorphism f is *hyperbolic* with respect to the set of handles H provided

(1) If $x \in h_i$, $f(x) \in h_j$ then $\text{int}(f(W_i^u(x))) \supset W_j^u(f(x))$ and $f(W_i^s(x)) \subset \text{int}(W_j^s(f(x)))$,

and

(2) if $x \in h_i$, $f(x) \in H$ and $v \in T_x(W_i^s(x))$, $w \in T_x(W_i^u(x))$ then $\|df(v)\| \leqslant \lambda \|v\|$ and $\|df(w)\| \geqslant \lambda^{-1} \|w\|$ for some $\lambda \in (0, 1)$ and $\| \ \|$ the standard metric on the handles $D_j^k \times D_j^{n-k}$.

From this, it is clear that $\Lambda = \bigcap_{n \in Z} f^n(H)$ has a hyperbolic structure and that if $x \in \Lambda$ then $W_i^s(x) \subset W^s(x)$ and $W_i^u(x) \subset W^u(x)$.

(3.11) DEFINITION. The *geometric intersection matrix* G corresponding to f and the hyperbolic handles set H is defined by $G_{ij} =$ number of components of $h_i \cap f(h_j)$.

Alternatively G_{ij} is the number of points of intersection of $W_i^s(y)$ with $f(W_j^u(x))$ for any $x \in h_j$, $y \in h_i$.

The following theorem which is the basis of much of what we do subsequently in these lectures was first proved by Smale [S4]. In the same article he proved some stability properties for diffeomorphisms with hyperbolic handles.

(3.12) THEOREM (HYPERBOLIC HANDLES). *If $f: M \longrightarrow M$ is a diffeomorphism which is hyperbolic with respect to a set of handles H and $\Lambda = \bigcap_{n \in Z} f^n(H)$, then $f|\Lambda$ is topologically conjugate to the subshift of finite type $\sigma(G): \Sigma_G \longrightarrow \Sigma_G$ where G is the geometric intersection matrix corresponding to f and H.*

PROOF. Suppose that H consists of m handles so that G is an $m \times m$ matrix. Let S be the set of all components of $H \cap f(H)$. We give the finite set S the discrete topology and define r, $l: S \longrightarrow \{1, 2, \ldots, m\}$ as follows. Let $S_{ij} \subset S$ be the set of components of $h_i \cap f(h_j)$, so that $G_{ij} =$ cardinality of S_{ij}, and define $l(s) = i$, $r(s) = j$ whenever $s \in S_{ij}$. If we define $s \longrightarrow s'$ whenever $r(s) = l(s')$ then clearly the subshift of finite type $\sigma: \Sigma \longrightarrow \Sigma$ determined by S and $\longrightarrow$ is just the edge shift of the matrix G.

We now define a map $\Psi: \Lambda \longrightarrow \Sigma$ by $\Psi(x) = s \in \Sigma$ where $s = (\ldots, s_{-1}, s_0, s_1, \ldots)$ and for all n, s_n is the element of S containing $f^{-n}(x)$, or equivalently $x \in f^n(s_n)$. Since $\Sigma \subset \Pi_{-\infty}^\infty S$ inherits the product topology it is clear that Ψ is continuous. Also it is immediate that $\Psi \circ f = \sigma \circ \Psi$. Since Λ is compact to complete the proof we need only show Ψ is a bijection, which we do by producing an inverse.

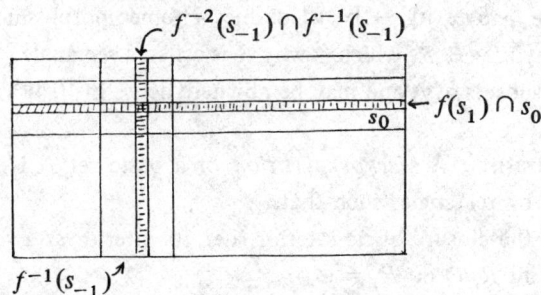

$$f^{-2}(s_{-1}) \cap f^{-1}(s_{-1})$$

$$\leftarrow f(s_1) \cap s_0$$

$$f^{-1}(s_{-1})$$

FIGURE (3.2)

Suppose $s = (\ldots, s_{-1}, s_0, s_1, \ldots) \in \Sigma$. Then $s_0 \cap f(s_1)$ is a single "strip" across h_i in the W^u direction where $i = l(s_0)$ (see Figure 3.2). This is because if $y \in h_i$ and $x \in s_1 \subset h_j$, then $f(W_j^u(x))$ intersects $W_i^s(y)$ in G_{ij} points, one point in each of the components in S_{ij}, and consequently one point in s_0. The fact that f is uniformly contracting on each $W_k^s(y)$ by a factor of λ means that in the W^s direction s_0 has diameter $\leqslant \lambda d$ and $s_0 \cap f(s_1)$ has diameter $\leqslant \lambda^2 d$ where d is the diameter of the D^k's. Of course in W^u direction the diameter of $s_0 \cap f(s_1)$ is d.

An induction on n shows that $K_n^+ = \bigcap_{k=0}^n f^k(s_k)$ crosses h_i in the W^u direction and has diameter in the W^s direction $\leqslant \lambda^n d$. A similar argument shows $K_n^- = \bigcap_{k=-n}^{-1} f^k(s_k)$ crosses h_i in the W^s direction and has diameter in the W^u direction $\leqslant \lambda^n d$. Hence $K_n = K_n^- \cap K_n^+ = \bigcap_{k=-n}^n f^k(s_k)$ is a compact set whose diameter is $\leqslant 2\lambda^n d$, which tends to zero as $n \longrightarrow \infty$. Therefore the nested sequence of nonempty compact sets K_n has an intersection $\bigcap_{n \geqslant 0} K_n$ which consists of a single point x. Thus $x = \bigcap_{n=-\infty}^\infty f^n(s_n)$, so the function $s \longrightarrow x$ is an inverse to Ψ. Since Ψ is a continuous bijection from Λ to Σ it is a homeomorphism.   Q.E.D.

We remark that Λ may not be a basic set for f, and, in fact, it may not be isolated in R. There can, possibly, be periodic points of f arbitrarily close to Λ. In our example at the beginning of this section this possibility is excluded because every point whose entire orbit does not remain in $h_1 \cup h_2$ tends to one of the other basic sets in either forward or backward time. In general we need similar information about f outside of H to claim Λ is isolated in R. To obtain the conclusion of the hyperbolic handles theorem (3.12) it is not even necessary that f be defined on all of M. We need only that f and f^{-1} are defined on H.

This result has been greatly generalized by Bowen [B1] who showed that all basic sets are closely related to subshifts of finite type and that on any basic set of dimension zero f is topologically conjugate to a subshift of finite type. We briefly describe his results.

The analogues of our handles are "rectangles" in the basic set Λ, which we now define. If $x \in \Lambda$ let $W_\epsilon^s(x) = \{y \in W^s(x) | d(f^n x, f^n y) \leqslant \epsilon \text{ for all } n \geqslant 0\}$, so $W_\epsilon^s(x)$ is a small disk centered at x in $W^s(x)$, and define $W_\epsilon^u(x)$ similarly. Since E_x^s is tangent to $W^s(x)$ at x and E_y^u is tangent to $W^u(y)$, it is easy to see that there is a $\delta > 0$ such that whenever $d(x, y) < \delta$, $W_\epsilon^s(x) \cap W_\epsilon^u(y)$ consists of a single point z. Using a Lyapunov function g, it is clear that $g(f^n(z)) \longrightarrow g(\Lambda)$ as $n \longrightarrow \infty$ and as $n \longrightarrow -\infty$ so $g(f^n(z)) = g(\Lambda)$ for all n, so $z \in \Lambda$. We denote the point $z = W_\epsilon^s(x) \cap W_\epsilon^u(y)$ by $[x, y]$. A subset R of Λ is called a *rectangle* if its diameter is less than δ and $[x, y] \in R$ whenever $x, y \in R$. Let $W^u(x, R) = R \cap W_\epsilon^u(x)$ and $W^s(x, R) = R \cap W_\epsilon^s(x)$; then one can show that the map $(x, y) \longrightarrow [x, y]$ defines a homeomorphism from $W^s(z, R) \times W^u(w, R)$ to R for any $z, w \in R$, which is why R is called a rectangle. Notice that R, $W^s(x, R)$ and $W^u(x, R)$ are all subsets of Λ and may be nowhere dense in M, $W^s(x)$ and $W^u(x)$ respectively.

(3.13) DEFINITION. A *Markov partition* of a basic set Λ is finite covering $\{R_1, \ldots, R_m\}$ of Λ by rectangles such that

(a) Each R_i is the closure of its interior (i.e., its interior as a subset of Λ).

(b) For $i \neq j$, int $R_i \cap$ int $R_j = \emptyset$.

(c) If $x \in$ int R_i, $f(x) \in$ int R_j then $f(W^u(x, R_i)) \supset W^u(f(x), R_j)$ and $f(W^s(x, R_i)) \subset W^s(f(x), R_j)$.

For example if the Λ of the hyperbolic handles theorem (3.12) is a basic set and we define $R_i = h_i \cap \Lambda$ then $\{R_i\}$ is a Markov partition. The following result of Bowen shows the intimate relationship between symbolic dynamics and the restriction of f to a basic set.

(3.14) THEOREM [B1]. *Suppose $f: M \to M$ is a diffeomorphism with hyperbolic chain recurrent set and Λ is a basic set of f. Then there exists a subshift of finite type σ: $\Sigma \to \Sigma$ and a continuous surjection $h: \Sigma \to \Lambda$ such that*

(a) $h \circ \sigma = f \circ h$.

(b) *If σ is a shift on the set of symbols S and for each $s \in S$ we define*

$$R_s = h(\{c = (\ldots, c_{-1}, c_0, c_1, \ldots) \in \Sigma \mid c_0 = s\})$$

then $\{R_s \mid s \in S\}$ forms a Markov partition of Λ.

(c) *There is a finite bound independent of x on the number of points in $h^{-1}(x)$, $x \in \Lambda$. If x is in the residual set $\bigcap_{n \in Z} f^n(\bigcup_s \operatorname{int} R_s)$, i.e., if $f^n(x) \in \bigcup_s \operatorname{int} R_s$ for all n, then $h^{-1}(x)$ consists of a single point.*

In an appendix we give a proof in the special case that Λ is zero dimensional. With this added hypothesis h is a homeomorphism and hence a topological conjugacy. In general however this is not the case since Σ is always zero dimensional while Λ may have higher dimension. See [B2] or [S1] for examples.

Chapter 4. Smale diffeomorphisms

In this chapter we explore the relationship between the symbolic dynamics of certain diffeomorphisms and homological properties of the diffeomorphisms on the chain level. In many ways the results are analogous to the results (2.11) and (2.16) for Morse gradients.

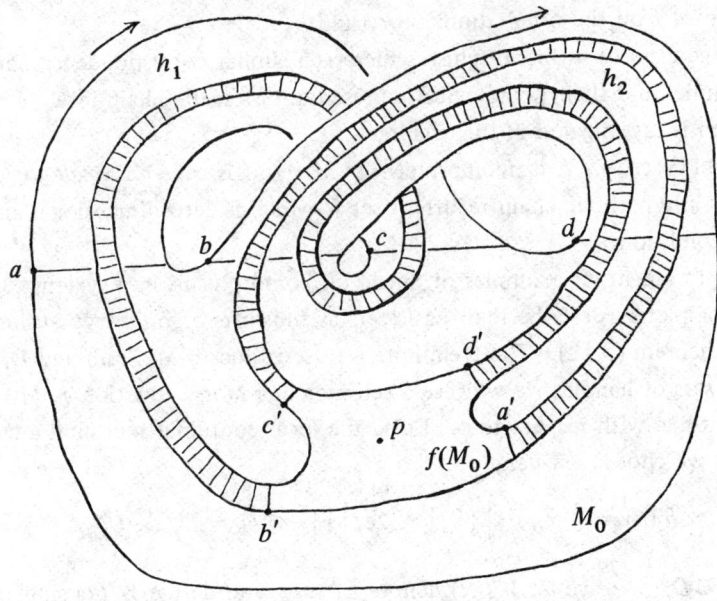

FIGURE (4.1)

As in the last chapter we begin with an example. Consider a diffeomorphism of the torus T^2 illustrated in Figure (4.1). What is shown is a picture of T^2 with a disk D^2 deleted and the image of $T^2 - D^2$ under a diffeomorphism $f: T^2 \longrightarrow T^2$. The diffeomorphism f will have an attracting fixed point p, a repelling fixed point ∞ in the missing D^2 and is constructed so that the two strips $H = h_1 \cup h_2$ are a set of hyperbolic handles for f (see (3.10)). The geometric intersection matrix for H with respect to f is $A = \begin{pmatrix} 0 & 1 \\ 1 & 1 \end{pmatrix}$. Thus if $\Lambda = \bigcap_{n=-\infty}^{\infty} f^n(H)$ then Λ is a compact invariant hyperbolic set and $f|\Lambda$ is topologically conjugate to $\sigma(A): \Sigma_A \longrightarrow \Sigma_A$.

If we let $M_0 \subset T^2$ be the closed set depicted in Figure (4.1) to which h_1 and h_2 are attached, and let $M_1 = M_0 \cup (h_1 \cup h_2)$ and $M_2 = T^2$, then it is clear from the construction that $\bigcap_{n \geqslant 0} f^n(M_0) = \{p\}$ and $\bigcap_{n \geqslant 0} f^{-n}(D_2) = \{\infty\}$ where $D_2 = \mathrm{cl}(M_2 - M_1)$. Since the chain recurrent set R is invariant it follows that $R \cap M_0 = \{p\}$, and $R \cap D_2 = \infty$, so $R \subset \{p, \infty\} \cup (\bigcap_{n-\infty}^{\infty} f^n(H)) = \{p, \infty\} \cup \Lambda$. Since the matrix A is irreducible, all of Λ is chain recurrent. Hence f has hyperbolic chain recurrent set and its basic sets are $\{p\}$, Λ, and $\{\infty\}$.

The matrix A is also related to a homological description of f in the following way. If we think of M_0 as a thickening of the point p, and the one-handles h_1 and h_2 as thickened 1-cells and D^2 as a 2-cell then the chain complex corresponding to this cell decomposition is given by $C_i = H_i(M_i, M_{i-1})$. If we assign orientations to h_1 and h_2 (actually orientations of $W_1^u(x)$ and $W_2^u(y)$ for some $x \in h_1$, $y \in h_2$) as in Figure (4.1) then h_1 and h_2 represent a basis of $C_1 = H_1(M_1, M_0)$ and the matrix of the map induced by f is $\widetilde{A} = \begin{pmatrix} 0 & 1 \\ -1 & 1 \end{pmatrix}$. Clearly this is just the geometric intersection matrix of f with respect to H with signs added to reflect the action of f on the orientations of h_1 and h_2.

For a large class of diffeomorphisms there is a similar correspondence between their symbolic dynamics and their homological representation on the chain level. It is this phenomenon which is investigated in this chapter.

(4.1) DEFINITION. A diffeomorphism $f: M \rightarrow M$ is called a *Smale diffeomorphism* provided it has a hyperbolic chain recurrent set R which is zero dimensional and satisfies the transversality condition.

We want to construct examples of Smale diffeomorphisms in a systematic way which allows us to completely describe their basic sets as subshifts of finite type using the hyperbolic handles theorem (3.12). The technique we use is due to M. Shub and D. Sullivan [SS]. To obtain the sets of handles we will use a self-indexing Morse function $\phi: M \rightarrow R$ and a flow g_t gradient-like with respect to ϕ. Using the local coordinates around a critical point p_i^k of index k, we choose a k-handle

$$h_i(k) = \{(\vec{x}, \vec{y}) \mid |\vec{x}|^2 \leqslant 2\epsilon, |\vec{y}|^2 \leqslant \epsilon\} = D_i^k \times D_k^{n-k}.$$

As before if $x \in D_i^k \times p$ we let $W_i^u(x)$ denote $D_i^k \times p$ and define $W_i^s(x)$ similarly.

(4.2) DEFINITION. A diffeomorphism $f: M \rightarrow M$ is said to be *fitted* with respect to the handle sets $H(k) = \bigcup h_i(k)$ provided

(a) For each k the diffeomorphism is hyperbolic with respect to the handle set $H(k)$.

(b) If $k < l$, $x \in h_i(l)$, and $y = f^n(x) \in h_j(k)$, for some $n \geqslant 0$ then

$$f^n(W_i^u(x)) \supset W_j^u(y).$$

Likewise,

$$f^{-n}(W_j^s(y)) \supset W_i^s(x).$$

(c) If $M_{k-1} = \phi^{-1}((-\infty, k - \epsilon])$ then

then $f(M_k) \subset \text{int}(M_{k-1} \cup H(k))$ for all $k \geqslant 0$, where ϕ is the self-indexing Morse function used to define the handle sets.

Recall that according to (3.12) it follows from (a) that f restricted to $\Lambda(k) = \bigcap_{n \in Z} f^n(H(k))$ is topologically conjugate to the shift $\sigma(G(k))$ where $G(k)$ is the geometric intersection matrix defined by $G(k)_{ij}$ = the number of components of $h_i(k) \cap f(h_j(k))$. We will show that the chain recurrent set is contained in $\bigcup_k \Lambda(k)$.

Thus using the algorithm from the proof of (3.7), for any fitted diffeomorphism we can explicitly describe the map on all of its basic sets in terms of the geometric intersection matrices $G(k)$.

(4.3) THEOREM. *If f is fitted with respect to $\{H(k)\}$ then f is a Smale diffeomorphism and on the basic sets of index k, f is topologically conjugate to the shift $\sigma(G(k))$ restricted to its chain recurrent set, where $G(k)$ is the geometric intersection matrix of f with respect to $H(k)$.*

PROOF. Let δ denote the minimum of the distances from $f(M_k)$ to $M - (M_{k-1} \cup H(k))$. By (c), $\delta > 0$. Suppose now $x \in R(f)$ is in $M_k - M_{k-1}$; then $f(x) \in f(M_k)$ so no chain starting at x and with jumps of length $< \delta$ can reach $M - (M_{k-1} \cup H(k))$. Since x is chain recurrent and $x \in M_k - M_{k-1}$ we must have $x \in H(k)$. Likewise $f(x)$ must be in $H(k)$ since $f(x) \in R$ and if $f(x) \notin M_k - M_{k-1}$ then the same argument shows $f(x) \in H(j)$ for some $j < k$ and then no δ chain could start at x and return to it. Thus if $x \in R \cap (M_k - M_{k-1})$, $f^n(x) \in H(k)$ for $n \geq 0$, and since $f^{-1}(R) = R$ we also have $f^n(x) \in H(k)$ for $n \leq 0$.

Hence if $\Lambda(k) = \bigcap_{n \in Z} f^n(H(k))$, we have $R \subset \bigcup_k \Lambda(k)$ and by the hyperbolic handles theorem (3.12) $f \mid \Lambda(k)$ is topologically conjugate to $\sigma(G(k))$. Clearly all basic sets of index k are contained in $\Lambda(k)$. The proof of (3.7) gives an algorithm for determining the basic sets of index k up to topological conjugacy from the matrix $G(k)$.

To show f is a Smale diffeomorphism we need also to check the transversality condition. But this is immediate from (b) of the definition of fitted since if $x \in \Lambda(k)$, $y \in \Lambda(l)$, $l < k$ and $z \in W^u(x) \cap W^s(y)$ then for some m, $f^m(z) \in h_j(l)$ and $f^{-m}(z) \in h_i(k)$ so $f^{2m}(W_i^u(f^{-m}(z))) \supset W_j^u(f^m(z))$ and hence is transverse to $W_j^s(f^m(y))$ at $f^m(z)$. Thus $W^u(f^m(x))$ is transverse to $W^s(f^m(y))$ at $f^m(z)$ and applying f^{-m} shows $W^u(x)$ intersects $W^s(y)$ transversely at z.   Q.E.D.

In the example with which we began this chapter we showed that if the matrix A was altered to include signs reflecting the action of f on orientations of the handles we obtained a chain level homological description of f. We want to duplicate this result for fitted diffeomorphisms. An *orientation* of a k-handle $h_i(k)$ will be the choice of an orientation for the k-dimensional disk $W_i^u(x)$ for any $x \in h_i(k)$. This of course determines an orientation for $W_i^u(y)$ for all $y \in h_i(k)$. Equivalently we could define an orientation of $h_i(k)$ to be an orientation of the normal bundle of $W_i^s(x)$.

Recall that if $P, Q \subset M$ are submanifolds with $\dim P + \dim Q = \dim M$ which intersect transversally at a point x and there are orientations assigned to P and the normal bundle of Q, then there is a well-defined *intersection number* $P \cdot Q(x)$ at x. It is defined to be 1 if the orientations of TP_x and NQ_x agree (where we have chosen a metric so the normal space to Q at x, NQ_x, equals TP_x), and -1 if they disagree. If P and Q intersect transversally at several points the total intersection number $P \cdot Q$ is defined to be $\Sigma_{x \in P \cap Q} P \cdot Q(x)$.

(4.4) DEFINITION. The *algebraic intersection matrix* $A(k)$ for the set of oriented handles $H(k)$ is defined by

$$A(k)_{ij} = f(W_j^u(p)) \cdot W_i^s(q),$$

the intersection number of $f(W_j^u(p))$ and $W_i^s(q)$, where $p \in h_j$, $q \in h_i$.

Recall that in the geometric intersection matrix $G(k)$, the ijth entry $G(k)_{ij}$ is simply the number of points of intersection of $f(W_j^u(p))$ with $W_i^s(q)$. The algebraic intersection matrix counts the points of intersection with a sign reflecting whether or not the orientation of $f(W_j^u(p))$ agrees with that of $W_i^u(q)$.

Recall from (2.11) that if $C_k = H_k(M_k, M_{k-1})$ and $\partial_k: C_k \longrightarrow C_{k-1}$ is the boundary map of the triple (M_k, M_{k-1}, M_{k-2}) then $C = \{C_k, \partial_k\}$ is a chain complex whose homology is $H_*(M)$. Also we showed that if p_i^k is the center of the handle $h_i(k)$ and $[W^u(p_i^k)]$ denotes the image in C_k of the generator of $H_k(W^u(p_i^k), W^u(p_i^k) \cap M_{k-1})$ corresponding to the orientation of $h_i(k)$ then $\{[W^u(p_i^k)]\}$ form a basis of C_k. Since for any $x \in h_i(k)$ the disk $W_i^u(x)$ is clearly homologous in (M_k, M_{k-1}) to $W_i^u(p_i^k)$ we will simply write $\{[W_i^u(k)]\}$ for this basis, and $W_i^u(k)$ for $W_i^u(p_i^k)$.

(4.5) PROPOSITION. *Let C be the chain complex for M given by $C_k = H_k(M_k, M_{k-1})$ and let $\tau_k: C_k \longrightarrow C_k$ be the chain map given by $\tau_k = f_{*k}: H_k(M_k, M_{k-1}) \longrightarrow H_k(M_k, M_{k-1})$, so $\tau_*: H_*(C) \longrightarrow H_*(C)$ is the same as $f_*: H_*(M) \longrightarrow H_*(M)$. Then the matrix of τ_k with respect to the basis $\{[W_i^u(k)]\}$ is the algebraic intersection matrix $A(k)$.*

PROOF.

$$\tau_k([W_j^u(k)]) = [f(W_j^u(p_j^k))] = \sum_i \sum_x f(W_j^u(k)) \cdot W_i^s(k)(x)[W_i^u(x)]$$

where the second sum is over the set of x in $f(W_j^u(k)) \cap W_i^s(k)$ and $\cdot$ denotes intersection number. Since $[W_i^u(x)] = [W_i^u(k)]$ we have $\tau_k([W_j^u(k)]) = \Sigma_i A_{ij}(k)[W_i^u(k)]$ which is the desired result.   Q.E.D.

Our long range goal is to understand what kind of (fitted) Smale diffeomorphisms can occur in a given homotopy class. In particular, we want to know what are the possibilities for the geometric intersection matrices $\{G(k)\}$ since they give a complete picture of the basic sets up to topological conjugacy.

We now give two theorems of Shub and Sullivan [SS], which represent the first steps in answering these questions and form the foundation on which later results are based.

(4.6) THEOREM [SS]. *If $f_0: M \longrightarrow M$ is a diffeomorphism of a compact manifold then f_0 is isotopic to a fitted diffeomorphism f. Moreover f can be chosen arbitrarily close to f_0 in the C^0 topology. If M has boundary $f: M \longrightarrow \text{int } M$ is an embedding.*

It follows that fitted Smale diffeomorphisms are dense in $\text{Diff}(M)$ with the C^0 topology.

If A is a matrix we denote by $|A|$ the matrix whose ijth entry is $|A_{ij}|$.

(4.7) THEOREM [SS]. *Suppose M is simply connected, has torsion free homology and dimension $n > 5$, and suppose $C = \{C_k, \partial_k\}_{k=0}^n$ is a finitely generated free chain complex with $H_*(C) = H_*(M)$ and $C_1 = C_{n-1} = 0$. If $f: M \rightarrow M$ is a diffeomorphism and $\tau = \{\tau_k: C_k \rightarrow C_k\}$ a chain map with τ_* on $H_*(C) = H_*(M)$ the same as the map f_* induced by f, then given any choice of bases for C_k, f is isotopic to a fitted diffeomorphism f_1, whose geometric intersection matrix $G(k) = |A(k)|$, for all k, where $A(k)$ is the matrix of τ_k with respect to the given basis (and also the algebraic intersection matrix).*

This theorem tells us a great deal about realizing geometric intersection matrices in fitted diffeomorphisms. It can be augmented by the following observation.

(4.8) PROPOSITION [F1]. *Suppose that $f_0: M \rightarrow M$ is a fitted diffeomorphism of a connected manifold of dimension $m \geq 3$ and that the handle sets H_0 and H_m consist of a single 0-handle and single m-handle respectively. If P is a nonnegative integer matrix the same size as the geometric intersection matrix $G(k)$, and $1 \leq k < m$, then f_0 is isotopic to a diffeomorphism f fitted with respect to the same handle sets and with geometric intersection matrices $G(j, f) = G(j, f_0)$, $j \neq k$, but $G(k, f) = G(k, f_0) + 2P$.*

The proofs of (4.6)–(4.8) have been relegated to Appendix B since they are somewhat lengthy and technical.

A somewhat weaker version of (4.6) was first proved by Smale [S5]. It said f is isotopic to a diffeomorphism with hyperbolic chain recurrent set and zero dimensional basic sets. Shub, Smale, and Williams added the strong transversality condition to this. The version of the theorem given here appeared subsequently in [SS]. Generalizations of (4.7) and several of the results of Appendix B for the case of nonsimply connected manifolds can be found in [Ma1].

For fitted diffeomorphisms Theorems (4.5) and (4.7) show the close connection between the dynamics on the basic sets and homological invariants. It is important to try to understand this connection when f is not fitted.

(4.9) DEFINITION. If $f: M \rightarrow M$ has a hyperbolic chain recurrent set and its basic sets are $\{\Omega_i\}_{i=0}^n$, then a *filtration associated to f* is a collection of submanifolds $M_0 \subset M_1 \subset \cdots \subset M_n = M$ such that
    (a) $f(M_i) \subset \text{int } M_i$ and
    (b) $\Omega_i = \bigcap_{n=-\infty}^{\infty} f^n(M_i - M_{i-1})$.

The existence of a filtration associated to f follows easily from the existence of a smooth Lyapunov function $\phi: M \rightarrow R$ (see (1.2)). We choose regular values c_i of ϕ such that $c_{i-1} < \phi(\Omega_i) < c_i$ (this is possible by (1.14)) and define $M_i = \phi^{-1}((-\infty, c_i])$. It is then clear that (a) holds. For (b) we note it is easy to show that if $x \in f^n(M_i - M_{i-1})$ for all n, $\lim_{n \to \infty} \phi(f^n(x)) = \phi(\Omega_i)$ and $\lim_{n \to \infty} \phi(f^{-n}(x)) = \phi(\Omega_i)$. Hence $\phi(f^n(x)) = \phi(\Omega_i)$ for all n so $x \in \Omega_i$.

If Ω_i is a zero dimensional basic set we would like to relate its dynamics to the action of f_* on $H_*(M_i, M_{i-1})$. We know from (3.14) or (A.5) that f restricted to a zero dimensional basic set is topologically conjugate to a subshift of finite type, $\sigma(G)$: $\Sigma_G \longrightarrow \Sigma_G$ for some matrix G. We want to alter the matrix G so that it includes more information, namely how Df acts on the bundle E^u.

Since Ω_i is zero dimensional the bundles $E^u(\Omega_i)$ and $E^s(\Omega_i)$ are trivial and we can choose orientations for them. We then define

$$\Delta(x) = \begin{cases} 1 & \text{if } Df_x: E_x^u \longrightarrow E_{f(x)}^u \text{ preserves orientation,} \\ -1 & \text{if } Df_x: E_x^u \longrightarrow E_{f(x)}^u \text{ reverses orientation.} \end{cases}$$

The function $\Delta(x)$ is locally constant. We choose a vertex shift $\sigma(G)$, representing $f|\Omega_i$, so G is a matrix of zeroes and ones, and Σ_G is the set of bi-infinite sequences in which j can follow i if and only if $G_{ij} = 1$. From the proof of (A.5), or directly, it is easy to see that we can choose G and the topological conjugacy h: $\Sigma_G \longrightarrow \Omega_i$ so that $\Delta(x)$ is constant on $h(C(j))$ where $C(j)$ is the cylinder set $\{c \in \Sigma_G | c_0 = j\}$. Let Δ_k be the value of $\Delta(x)$ on $h(C(k))$.

(4.10) DEFINITION. The matrix A defined by $A_{jk} = \Delta_k G_{jk}$ will be called a *structure matrix* for the basic set Ω_i.

Since $G = |A|$, a structure matrix for Ω_i contains a description of $f|\Omega_i$ up to topological conjugacy. But A contains the additional information of how Df acts on the orientation of $E^u(\Omega_i)$.

We now give a result relating structure matrices to homological invariants of f and an associated filtration. For an endomorphism of a vector space $e: V \longrightarrow V$ we will denote by e^+ the *nonnilpotent part* of e, i.e., the map induced by e on the quotient V/V_0 where $V_0 = \{v \in V | e^k(v) = 0 \text{ for some } k > 0\}$. Alternatively e^+ is the restriction of e to the subspace of V spanned by the generalized eigenspaces of e corresponding to nonzero eigenvalues. Two endomorphisms e_i: $V_i \longrightarrow V_i$, $i = 1, 2$, are said to be *conjugate* if there is an isomorphism Ψ: $V_1 \longrightarrow V_2$ such that $e_2 \circ \Psi = \Psi \circ e_1$.

(4.11) THEOREM [BF]. *Suppose f: $M \longrightarrow M$ has a hyperbolic chain recurrent set, $\{M_j\}$ is an associated filtration and Ω_i is a zero dimensional basic set of index u. If A is an $n \times n$ structure matrix for Ω_i, and F is a field, then*

(a) *The nonnilpotent part A^+ of A: $F^n \longrightarrow F^n$ is conjugate to the nonnilpotent part f_{*u}^+ of f_{*u}: $H_u(M_i, M_{i-1}; F) \longrightarrow H_u(M_i, M_{i-1}; F)$, and*

(b) *the map f_{*k}: $H_k(M_i, M_{i-1}; F) \longrightarrow H_k(M_i, M_{i-1}; F)$ is nilpotent for all $k \neq u$.*

We will prove a generalization of (b) in Chapter 6. The proof of (a) can be found in (3.2) and (3.3) of [BF]. We point out that the structure matrix (and even its size) is non-unique. Likewise, there are many possible choices for $\{M_j\}$ and dim $H_u(M_i, M_{i-1}; F)$ can vary considerably. But in both cases all the lack of uniqueness affects only the nilpotent part of the induced endomorphisms.

Chapter 5. Zeta functions

In the previous chapter we developed a systematic way of constructing examples of diffeomorphisms with hyperbolic chain recurrent set. The theorems there give sufficient conditions for the existence of diffeomorphisms in a given homotopy class with prescribed basic sets. In this chapter we develop invariants which can represent obstructions to the existence of diffeomorphisms with prescribed basic sets in some homotopy classes. Ultimately one hopes that the necessary conditions given by these invariants will converge with sufficient conditions obtained by explicit construction of examples.

The first invariant is the zeta function for diffeomorphisms introduced by Artin and Mazur [AM].

(5.1) DEFINITION. The *zeta function* of a self-map $f: M \longrightarrow M$ is the formal power series in t

$$\zeta(f) = \exp\left(\sum_{m=1}^{\infty} \frac{1}{m} N_m t^m \right)$$

where N_m is the cardinality of the fixed point set of f^m. It is defined if $N_m < \infty$ for all m.

It turns out that this is often a rational function; for example, when f has hyperbolic chain recurrent set (see (5.18) below). In this case a finite set of complex numbers, the zeroes and poles of ζ, determine all of the numbers N_m, which are invariants of topological conjugacy.

The choice of the series defining $\zeta(f)$ goes back to work of A. Weil in algebraic geometry. The purpose of this formalism in its simplest version is to express the nonzero eigenvalues of a matrix A in terms of the traces of its powers.

(5.2) LEMMA. *If A is an $n \times n$ complex matrix then*

$$\exp\left(\sum_{m=1}^{\infty} \frac{1}{m} (\text{tr } A^m) t^m \right) = \frac{1}{\det(I - At)}.$$

PROOF. The series $\sum_{m=1}^{\infty} A^m t^m / m$ is the formal power series for $-\log(I - At)$, which will converge for t near zero. It is a well-known fact that, for any matrix B, $\exp(\text{tr } B) = \det \exp(B)$ (this is easy for diagonalizable matrices which are dense in all matrices and hence it follows in general by continuity).

Thus

$$\exp\left(\sum \frac{1}{m}(\mathrm{tr}\ A^m)t^m\right) = \exp\left(\mathrm{tr}\sum \frac{1}{m}A^m t^m\right) = \exp(\mathrm{tr}(-\log(I-At)))$$

$$= \det(\exp(-\log(I-At))) = 1/\det(I-At). \quad \text{Q.E.D.}$$

It is useful to note that the roots of $\det(I-At)$ are the reciprocals of the nonzero eigenvalues of A and they will have the same multiplicity as the corresponding eigenvalues. The polynomial $\det(I-At)$ contains the same information as the characteristic polynomial of A except the multiplicity of zero as an eigenvalue, if it is one, has been lost.

We can now calculate the zeta function of a subshift of finite type. This was first done by Bowen and Lanford [**BL**].

(5.3) COROLLARY [**BL**]. *If $\sigma(A): \Sigma_A \rightarrow \Sigma_A$ is the subshift of finite type associated to the nonnegative integer matrix A, then the zeta function of $\sigma(A)$ is given by*

$$\zeta(\sigma) = 1/\det(I-At).$$

PROOF. According to (3.4) the cardinality of the fixed point set of $\sigma(A)^m$ is given by $N_m = \mathrm{tr}\ A^m$. Now (5.2) gives the desired result. Q.E.D.

Much of the value of the zeta function as an invariant stems from its relationship with homological invariants of f. This relationship arises via the Lefschetz fixed point theorem which we now discuss. Our treatment of the fixed point index is based on that of Dold [**D2**]. The fixed point index is defined for self-maps of Euclidean neighborhood retracts (ENR's). These are spaces which can be embedded in some R^n in such a way that the image of the embedding is a retraction of a neighborhood of itself, i.e., X is an ENR if there is an open set $U \subset R^n$ and maps $i: X \rightarrow U$, $r: U \rightarrow X$ such that $r \circ i(x) = x$ for all $x \in X$.

(5.4) DEFINITION. If $V \subset R^n$ is open and $f: V \rightarrow R^n$ has a compact set F of fixed points then the *fixed point index* $I_F(f)$ is defined to be the integer $(\mathrm{id} - f)_*(u_F) \in H_n(R^n, R^n - \{0\}) = Z$. The map $(\mathrm{id} - f): (V, V - F) \rightarrow (R^n, R^n - \{0\})$ is given by $(\mathrm{id} - f)(x) = x - f(x)$ and $u_F \in H_n(V, V - F)$ is the image of 1 under the composite $Z = H_n(R^n, R^n - D^n) \rightarrow H_n(R^n, R^n - F) \cong H_n(V, V - F)$ where D^n is an n-disk containing F.

Thus, for example if $p \in R^n$ is an isolated fixed point of a map $f: R^n \rightarrow R^n$, $I_p(f) = (\mathrm{id} - f)_*(u_p)$ where $u_p \in H_n(R^n, R^n - \{p\})$ is the standard generator.

Suppose now that E is an ENR with embedding $i: E \rightarrow U$, U an open set in R^n, and retraction $r: U \rightarrow E$.

(5.5) DEFINITION. If $V \subset E$ is an open set and the fixed point set F of $f: V \rightarrow E$ is compact, then the *fixed point index* is defined by

$$I_F(f) = I_F(i \circ f \circ r) \quad \text{where } i \circ f \circ r: r^{-1}(V) \rightarrow R^n.$$

For a proof that $I_F(f)$ is independent of the embedding $i: E \rightarrow U \subset R^n$ and the retraction $r: U \rightarrow E$, see (2.3) of [**D2**].

We will be primarily interested in hyperbolic fixed points of a smooth map $f: M \rightarrow M$.

(5.6) DEFINITION. A point $x \in M$ is a *hyperbolic periodic point* of period n provided $f^n(x) = x$, $f^i(x) \neq x$ for $1 \leqslant i < n$ and $Df_x^n : TM_x \longrightarrow TM_x$ has no eigenvalues of absolute value one. The *index* u of p is the dimension of the subspace $E_x^u \subset TM_x$ spanned by generalized eigenspaces with eigenvalues > 1 in absolute value.

This definition is equivalent to saying that the orbit of x is a finite invariant set with a hyperbolic structure in the sense of (1.6), and then E_x^u is the fiber of E^u over x.

(5.7) PROPOSITION. *Suppose p is the only fixed point of a smooth map $f : V \longrightarrow M$ where $V \subset M$ is open and $\det(I - Df_p) \neq 0$; then*

$$I_p(f) = \mathrm{sgn}(\det(I - Df_p)).$$

If p is hyperbolic then $I_p(f) = (-1)^u \Delta$, where $u = \dim E_p^u$ is the index of p and $\Delta = 1$ if $Df_p : E_p^u \longrightarrow E_p^u$ preserves orientation and -1 otherwise.

PROOF. Suppose first that $M = R^n$; then the fact that $\det(I - Df_p) \neq 0$ and the implicit function theorem imply that (id $- f$) is a local diffeomorphism at p which preserves or reverses orientation depending on whether or not $\det(I - Df_p)$ is positive or negative. Thus (id $- f)_* : H_n(V, V - \{p\}) \longrightarrow H_n(R^n, R^n - \{0\})$ is $\pm$id: $Z \longrightarrow Z$, the sign depending on the sign of $\det(I - Df_p)$, so $I_p(f) = \mathrm{sgn}(\det(I - Df_p))$.

We will show $I_p(f) = (-1)^u \Delta$, if p is hyperbolic, by showing $(-1)^u \Delta = \mathrm{sgn}(\det(I - Df_p))$. We do this by noting that if $\{\mu_i\}$ and $\{\lambda_j\}$ are the eigenvalues of Df_p with $|\mu_i| > 1$ and $|\lambda_j| < 1$ then $\det(I - Df_p) = \Pi(1 - \lambda_j)\Pi(1 - \mu_i)$. Complex eigenvalues contribute nothing to the sign of this since they come in conjugate pairs and $(1 - \lambda)(1 - \bar{\lambda}) = |1 - \lambda|^2 > 0$. Since a real linear map with complex eigenvalues always preserves orientation and $u = $ number of eigenvalues μ_i with $|\mu_i| > 1$, the addition or deletion of a complex eigenvalue also does not affect $(-1)^u \Delta$.

Hence we may assume Df_p has only real eigenvalues. Then it is clear $\mathrm{sgn}(\det(I - Df_p))$ $= \mathrm{sgn}(\Pi(1 - \mu_i)) = (-1)^{u-r}$, where $r = $ number of μ_i which are negative. But $\Delta = (-1)^r$ $= (-1)^{-r}$, so $\mathrm{sgn}(\det(I - Df_p)) = (-1)^u \Delta$.

This completes the proof when $M = R^n$. In general we use the definition of the fixed point index for ENR's (5.5) and choose an embedding $i : M \longrightarrow U \subset R^n$ and retraction $r : U \longrightarrow M$. Since $\det(I - D(i \circ f \circ r)_{i(p)}) = \det(I - Df_p)$, and it is clear that $(-1)^u \Delta$ is the same for f and $i \circ f \circ r$, the result follows.   Q.E.D.

We now consider the basic properties of the fixed point index, following the presentation of [D2].

(5.8) THEOREM. *Suppose $f : E \longrightarrow E$ is a continuous map of an ENR with compact fixed point set F.*

(a) *If $F_0 \subset F$ is compact and V, V' are open sets in E with $F_0 = V \cap F = V' \cap F$ then $I_{F_0}(f|V) = I_{F_0}(f|V')$. Hence we denote this $I_{F_0}(f)$.*

(b) *If $F = F_1 \cup F_2$ where F_1, F_2 are compact and disjoint then $I_F(f) = I_{F_1}(f) + I_{F_2}(f)$.*

(c) *If* $f_t: E \rightarrow E$, $t \in [0, 1]$ *is a homotopy and* $K = \{x \mid f_t(x) = x \text{ for some } t\}$ *is compact then* $I_{F_0}(f_0) = I_{F_1}(f_1)$ *where* F_0 *and* F_1 *are the fixed point sets of* f_0 *and* f_1 *respectively.*

PROOF. From the definition of the index we may as well assume E is an open set in R^n. Now if $f: E \rightarrow R^n$ has compact fixed point set F and we are given any compact K and open W satisfying $F \subset K \subset W \subset E$ it is easy to show from the definition that $I_F(f) = (\text{id} - f)_*(u_K) \in H_n(R^n, R^n - \{0\}) = Z$, where $(\text{id} - f): (W, W - K) \rightarrow (R^n, R^n - \{0\})$ and u_K is the image of 1 under the composite $Z = H_n(R^n, R^n - D^n) \rightarrow H_n(R^n, R^n - K) \cong H_n(W, W - K)$ and $D^n \supset K$ is an n-disk. This is because the inclusion $(W, W - K) \rightarrow (E, E - F)$ takes u_K to u_F. Applying this with $K = F_0$, $E = V \cup V'$ and $W = V$ or V' proves (a).

Now to prove (c) take $K = \{x \mid f_t(x) = x \text{ for some } t\}$ and $W = E$, so $I_{F_0}(f_0) = (\text{id} - f_0)_*(u_K) = (\text{id} - f_1)_*(u_K) = I_{F_1}(f_1)$, since f_0 and f_1 are homotopic.

To prove (b) choose disjoint neighborhoods V_1 and V_2 of F_1 and F_2. Now taking $W = V_1 \cup V_2$ and $K = F_1 \cup F_2$ our remarks above indicate $I_F(f) = I_F(f \mid W)$ which from the definition of the index equals $I_{F_1}(f \mid V_1) + I_{F_2}(f \mid V_2) = I_{F_1}(f) + I_{F_2}(f)$.   Q.E.D.

Before proving the Lefschetz fixed point theorem we need some algebraic facts. Suppose C is a finitely generated free abelian group or a finite dimensional vector space and $e: C \rightarrow C$ is an endomorphism. We will be interested in functions which assign to each such endomorphism an element of an abelian group (often Z) and which are additive on short exact sequences. That is, if α is such a function and

$$
\begin{array}{ccccccccc}
0 & \longrightarrow & C_1 & \longrightarrow & C_2 & \longrightarrow & C_3 & \longrightarrow & 0 \\
 & & \downarrow e_1 & & \downarrow e_2 & & \downarrow e_3 & & \\
0 & \longrightarrow & C_1 & \longrightarrow & C_2 & \longrightarrow & C_3 & \longrightarrow & 0
\end{array}
$$

is a commutative diagram with the horizontal rows exact then $\alpha(e_2) = \alpha(e_1) + \alpha(e_3)$. Such a function α is called an Euler-Poincaré function (see [L, p. 98]). An example is $\alpha(e) = \text{tr}(e) \in Z$. This is easily seen by choosing bases for C_1, C_2, C_3 and matrices for e_1, e_2, e_3.

(5.9) PROPOSITION. *Suppose* α *is a function which assigns to each pair* (C, e), *with* e *an endomorphism of* C, *an element of an abelian group* G. *If* α *is additive on short exact sequences then for any finitely generated free chain complex* $C = \{C_i\}$ *and chain map* $\tau = \{\tau_i: C_i \rightarrow C_i\}$ *we have*

$$\sum(-1)^i \alpha(C_i, \tau_i) = \sum(-1)^i \alpha(\tau_{i*}, H_i(C)).$$

A proof can be found in [L, p. 98].

We are now in a position to prove the Lefschetz fixed point theorem which will be our most important tool for relating dynamics to homology.

(5.10) THEOREM (LEFSCHETZ). *Suppose that* $f: E \rightarrow E$ *is a continuous map of a compact ENR and has fixed point set* F. *Then* $I_F(f) = \Sigma(-1)^i \text{tr}(f_{*i})$ *where* $f_{*i}: H_i(E, R) \rightarrow H_i(E, R)$ *is induced by* f.

PROOF. We first consider the case where E is a manifold and f is a diffeomorphism. Since the index is a homotopy invariant, by (4.6) we can assume that f is fitted. If Λ_k is $\bigcap_{n=-\infty}^{\infty} f^n(H(k))$ then $f|\Lambda_k$ is topologically conjugate to the subshift of finite type with matrix the geometric intersection matrix $G(k)$. All fixed points are in some Λ_k and the number in Λ_k is given by $\operatorname{tr} G(k)$ (see (3.4)). If we want to count fixed points, all of which are hyperbolic, with a sign representing their fixed point index we get by (5.7)

$$\sum_{p \in F(k)} I_p(f) = \sum_{p \in F(k)} (-1)^k \Delta_p$$

where $F(k) = F \cap \Lambda_k$ and $\Delta_p = \pm 1$ depending on the action of Df_p on the orientation of E_p^u. But if $p \in h_i(k)$, then Δ_p is the intersection number $f(W_i^u(p)) \cdot W_i^s(p)$ so it follows that $\Sigma_{p \in F(k)} \Delta_p = \operatorname{tr} A(k)$ where $A(k)$ is the algebraic intersection matrix for $H(k)$. Hence $\Sigma_{p \in F(k)} I_p(f) = (-1)^k \operatorname{tr} A(k)$. Therefore $I_F(f) = \Sigma_{p \in F} I_p(f) = \Sigma_{k=1}^n (-1)^k \operatorname{tr} A(k)$. By (4.4) $\{A(k)\}$ are matrices for chain maps representing f on a chain complex whose homology is $H_*(E)$. Thus by (5.9)

$$I_F(f) = \sum_{k=0}^n (-1)^k \operatorname{tr}(A(k)) = \sum_{k=1}^n (-1)^k \operatorname{tr}(f_{*k}).$$

This proves the result when f is a diffeomorphism. If $f: E \longrightarrow E$ is a self-embedding of a manifold with boundary the same proof works since f is still homotopic to a fitted self-embedding.

Now if $f: E \longrightarrow E$ is a continuous map of a compact ENR we note that we can replace E by a compact neighborhood M in R^n which is a manifold with boundary and extend f to M. If we then embed M in R^m, m large, we can choose a tubular neighborhood V of M and approximate $f: M \longrightarrow V$ by an embedding $f': M \longrightarrow V$. Finally extend f' to an embedding $\widetilde{f}: V \longrightarrow V$ by making it contract normal bundle fibers. Since $I_F(f) = I_F(\widetilde{f})$ and it is easy to check $\operatorname{tr} \widetilde{f}_{*k} = \operatorname{tr} f_{*k}$ the result follows from the fact that it holds for self-embeddings of manifolds. Q.E.D.

(5.11) COROLLARY. *Suppose $f: (X, A) \longrightarrow (X, A)$ is a continuous map of a compact pair of ENR's with fixed point set F. If $F_1 = F \cap (X - A)$, $F_2 = F \cap A$ and $F_2 \subset \operatorname{int} A$, then*

$$I_{F_1}(f) = \sum (-1)^k \operatorname{tr} f_{*k}$$

where $f_{*k}: H_k(X, A; R) \longrightarrow H_k(X, A; R)$ is induced by f.

PROOF. We consider the long exact sequence of the pair (X, A), with endomorphisms induced by f, $\cdots \longrightarrow H_k(A) \longrightarrow H_k(X) \longrightarrow H_k(X, A) \longrightarrow H_{k-1}(A) \longrightarrow \cdots$ as a chain complex with trivial homology. Applying (5.9) we get

$$\sum (-1)^k \operatorname{tr} f_{*k}(A) - \sum (-1)^k \operatorname{tr} f_{*k}(X) + \sum (-1)^k \operatorname{tr} f_{*k}(X, A) = 0.$$

Thus from (5.10) we get

$$I_{F_2}(f|A) - I_F(f) + \sum(-1)^k \operatorname{tr} f_{*k}(X, A) = 0.$$

It follows that $\Sigma(-1)^k \operatorname{tr} f_{*k}(X, A) = I_F(f) - I_{F_2}(f) = I_{F_1}(f)$.   Q.E.D.

(5.12) DEFINITION. Suppose $f: E \longrightarrow E$ is a map of a compact ENR and, for each $n > 0$, $F(n)$ is the set of fixed points of f^n, then the *homology zeta function* $Z(f)$ of f is defined by $Z(f) = \exp \Sigma_{n=1}^{\infty} L(f^n) t^n /n$ where $L(f^n) = I_{F(n)}(f^n)$ is called the *Lefschetz number of* f^n.

Thus instead of using the number of fixed points of f^n for a coefficient as we do in $\zeta(f)$ we use the sum of the fixed point indices $L(f^n) = \Sigma_{p \in F(n)} I_p(f^n)$. The Lefschetz fixed point theorem allows us to compute $Z(f)$ quite easily in terms of the action of f on homology. This makes $Z(f)$ much more useful than $\zeta(f)$ which is less easily computed.

(5.13) PROPOSITION. *If* $f: E \longrightarrow E$ *is a map of a compact ENR, then the homology zeta function of* f *is a rational function given by*

$$Z(f) = \prod_{k=0}^{n} \det(I - f_{*k}t)^{(-1)^{k+1}},$$

where $f_{*k}: H_k(E; R) \longrightarrow H_k(E; R)$ *is induced by* f, *and* n *satisfies* $H_k(E) = 0$ *for* $k > n$.

PROOF. By definition

$$Z(f) = \exp \sum_{m=1}^{\infty} L(f^m) t^m /m$$

and by (5.10)

$$L(f^m) = \sum_{k=0}^{n} (-1)^k \operatorname{tr} f_{*k}^m.$$

Thus

$$Z(f) = \exp \sum_{m=1}^{\infty} \left(\sum (-1)^k \operatorname{tr} f_{*k}^m \right) t^m /m$$

$$= \prod_{k=0}^{n} \left(\exp \sum_{m=1}^{\infty} \frac{1}{m} \operatorname{tr} f_{*k}^m t^m \right)^{(-1)^k}$$

$$= \prod_{k=0}^{n} \det(I - f_{*k}t)^{(-1)^{k+1}} \quad \text{by (5.2).   Q.E.D.}$$

If $\bar{f}: (X, A) \longrightarrow (X, A)$ is a map of a compact pair of ENR's with $\operatorname{Fix}(\bar{f}^n) = \{x \in X | \bar{f}^n(x) = x\}$ satisfying $(\operatorname{Fix}(\bar{f}^n) \cap A) \subset \operatorname{int} A$ for all n then we can define the homology zeta function of the map $\bar{f}$ of the pair by

$$Z(\bar{f}) = \exp \sum_{m=1}^{\infty} L(\bar{f}^m) t^m / m$$

where $L(\bar{f}^m) = I_{F(n)}(\bar{f}^m | X - A)$ and $F(n) = \text{Fix}(f^n) \cap (X - A)$. We then have the following analog of (5.13), whose proof is essentially identical.

(5.14) PROPOSITION. *If* $\bar{f} : (X, A) \rightarrow (X, A)$ *is a self-map of a compact ENR pair with* $\text{Fix}(\bar{f}^n) \cap A \subset \text{int } A$ *for all* n, *then the homology zeta function of* $\bar{f}$ *is given by*

$$Z(\bar{f}) = \prod_{k=1}^{n} \det(I - \bar{f}_{*k} t)^{(-1)^{k+1}}$$

where $\bar{f}_{*k} : H_k(X, A; R) \rightarrow H_k(X, A; R)$ *is induced by* $\bar{f}$ *and* n *satisfies* $H_k(X, A) = 0$ *if* $k > n$.

We return now to the consideration of diffeomorphisms with hyperbolic chain recurrent set. For such a diffeomorphism $f : M \rightarrow M$ with basic sets $\{\Omega_i\}_{i=0}^{m}$ we can choose a smooth Lyapunov function $\phi : M \rightarrow R$ (see (1.2)) and regular values of ϕ, $\{c_i\}$, such that $c_{i-1} < \phi(\Omega_i) < c_i$. We then define $M_i = \phi^{-1}((-\infty, c_i])$. This gives a filtration $M_0 \subset M_1 \subset \cdots \subset M_m = M$ with $\Omega_i = \bigcap_{n=-\infty}^{\infty} f^n(M_i - M_{i-1})$. Recall that a set of manifolds with boundary $\{M_i\}$ constructed in this way is called a filtration for f (see (4.9)).

We will be interested in analyzing the zeta function of f restricted to a single basic set.

(5.15) DEFINITION. If $f : M \rightarrow M$ has a hyperbolic chain recurrent set an basic sets $\{\Omega_i\}$ we denote by ζ_i the zeta function of $f | \Omega_i$ and by Z_i the homology zeta function of f restricted to Ω_i, i.e.,

$$Z_i(f) = \exp \sum_{m=1}^{\infty} \frac{1}{m} \widetilde{N}_m t^m$$

where $\widetilde{N}_m = \Sigma I_p(f^m)$, the sum being taken over all $p \in \text{Fix}(f^m) \cap \Omega_i$.

(5.16) PROPOSITION. *Suppose* $f : M \rightarrow M$ *has a hyperbolic chain recurrent set and basic sets* $\{\Omega_i\}_{i=0}^{l}$; *then*
 (a) $\zeta(f) = \Pi_{i=0}^{l} \zeta_i(f)$,
 (b) $Z(f) = \Pi_{i=0}^{l} Z_i(f)$,
 (c) $Z_i(f) = \Pi_{k=0}^{n} \det(I - f_{*k} t)^{(-1)^{k+1}}$
where $f_{*k} : H_k(M_i, M_{i-1}; R) \rightarrow H_k(M_i, M_{i-1}; R)$ *is induced by* f, *and* $\{M_i\}$ *is a filtration associated to* f.

PROOF. Since $\text{Fix}(f^m) = \bigcup_i \text{Fix}(f^m | \Omega_i)$, (a) follows from the definitions of $\zeta(t)$ and $\zeta_i(t)$. Likewise if $\widetilde{N}_m(i) = \Sigma_{p \in F(i)} I_p(f^m)$ where $F(i) = \text{Fix}(f^m) \cap \Omega_i$ then $\Sigma_{i=0}^{l} \widetilde{N}_m(i) = L(f^m)$ so (b) also follows from definitions. Finally (c) is an immediate consequence of (5.14). Q.E.D.

(5.17) PROPOSITION. *If* Ω_i *is a basic set of* $f : M \rightarrow M$ *and* $Df : E^u \rightarrow E^u$ *preserves (or reverses) an orientation of the bundle* E^u *over* Ω_i, *then*

$$\zeta_i(t) = Z_i(t)^{(-1)^u} \qquad (or \ Z_i(-t)^{(-1)^u})$$

where u is the index of Ω_i.

PROOF. If Df preserves orientation, N_m = cardinality of $\mathrm{Fix}(f^m) \cap \Omega_i$, and $\widetilde{N}_m = \Sigma I_p \ (f^m), p \in \mathrm{Fix}(f^m) \cap \Omega_i$ then $N_m = (-1)^u \widetilde{N}_m$ since $I_p(f^m) = (-1)^u$. Hence it follows that $\zeta_i(t) = Z_i(t)^{(-1)^u}$.

If Df reverses orientation then $I_p(f^m) = (-1)^u \Delta_p$ and $\Delta_p = (-1)^m$, so $\widetilde{N}_m = (-1)^u(-1)^m N_m$. Thus

$$Z_i(-t) = \exp\left(\sum \frac{1}{m} \widetilde{N}_m (-t)^m\right)$$

$$= \exp\left(\sum \frac{1}{m}(-1)^u N_m t^m\right) = \zeta_i(t)^{(-1)^u}. \quad \text{Q.E.D.}$$

In general however E^u may not be orientable, and even if it is Ω_i may not be connected so Df can preserve orientation on part of Ω_i and reverse it on the rest. This is in fact typical for fitted diffeomorphisms.

We are now prepared to prove the rationality of $\zeta(t)$ when $f: M \longrightarrow M$ has hyperbolic chain recurrent set. This was first proved by J. Guckenheimer [Gu] using a technique developed by R. F. Williams [Wm2]. Subsequently A. Manning proved a slightly more general theorem using Markov partitions. The proof we give is different from both of these but depends, as does the Guckenheimer-Williams proof, on utilizing the homology zeta function.

(5.18) THEOREM. *If* $f: M \longrightarrow M$ *has a hyperbolic chain recurrent set and basic sets* $\{\Omega_i\}$, *then for each i,* $\zeta_i(t)$ *is a rational function. In fact, it is a quotient of integer polynomials whose constant term is 1.*

PROOF. The idea of the proof is to produce a map of an ENR whose homology zeta function equals $\zeta_i(t)$. Since homology zeta functions are always rational this will suffice. Actually we produce a map $\bar{g}: (X, A) \longrightarrow (X, A)$ of a pair.

The bundle E^u can be extended to a bundle (also called E^u) over a neighborhood U_0 of Ω_i. We choose a filtration associated to f so that $\Omega_i = \bigcap_{n=-\infty}^{\infty} f^n(M_i - M_{i-1})$. It follows that for N large if $U = f^N(M_i) - f^{-N}(M_{i-1})$, then $U \cup f(U) \subset U_0$. Let $X_0 = (E^u|U) \cup T(f^{-N}(M_{i-1}))$ so X_0 is the collection of all tangent vectors over $f^{-N}(M_{i-1})$ together with all tangent vectors over U which are in E^u. We have a bundle map Df defined on $T(f^{-N}(M_{i-1}))$ and on $E^u|\Omega_i$ which we extend to all of X_0 as follows. Let $\alpha: M \longrightarrow [0, 1]$ be a smooth function which is 0 on a neighborhood of $f^{-N}(M_{i-1})$ and 1 on a neighborhood of $(M - f^{-N-1}(M_{i-1}))$. Define a map $g: X_0 \longrightarrow X_0$ by

$$g(v_x) = \begin{cases} Df(v_x) & \text{if } x \in f^{-N}(M_{i-1}), v_x \in TM_x, \\ \alpha(x)P_{f(x)}^u(Df(v_x)) + (1 - \alpha(x))Df(v_x) & \text{if } x \in U, v_x \in E_x^u, \end{cases}$$

where $P_x^u \colon TM_x \longrightarrow E_x^u$ is orthogonal projection in some Riemannian metric. If U was chosen small enough, the bundle $E^u|U$ is nearly enough invariant under Df that $v_x \neq 0$ implies $g(v_x) \neq 0$.

Now let X be the one point compactification of X_0, i.e., we add a single point ∞ which is "at infinity" in all the fibers TM_x, $x \in f^{-N}(M_{i-1})$ and E_x^u, $x \in U$. We define $A \subset X$ to be $T(f^{-N}(M_{i-1})) \cup \{v \in E_x^u | x \in U \text{ and } |v| \geqslant 1\} \cup \{\infty\}$. Clearly g extends to X by setting $g(\infty) = \infty$, and $g(A) \subset A$ since g expands the length of vectors $v_x \in E^u$, $x \in U$.

Let $\bar{g} \colon (X, A) \longrightarrow (X, A)$ be the map of the pair determined by g. The homology zeta function $Z(\bar{g})$ of this self-map of the pair is rational by (5.14). On the other hand it is defined to be $\exp \Sigma L(\bar{g}^m) t^m/m$ where $L(\bar{g}^m) = \Sigma I_p(g^m)$, $p \in \text{Fix}(g^m) \cap (X - A)$. The fixed points of g^m in $X - A$ all lie in Ω_i and are precisely the fixed points of f^m. To calculate $I_p(g)$ we use (5.7) noting that p is a hyperbolic fixed point of g on $E^u|U$. But the unstable space $E_p^u(g)$ is $E_p^u(f) \oplus E_p^u(f)'$, the sum of a $E_p^u(f)$ in TU_x and $E_p^u(f)'$, the tangent space to the fiber of $E^u|U$ at p. Also $Dg_p \colon E_p^u(g) \longrightarrow E_p^u(g)$ is the sum of two copies of Df_p: $E_p^u(f) \longrightarrow E_p^u(f)$. Hence $I_p(g) = (-1)^{2u(i)} \Delta$ by (5.7), but $\Delta = 1$ since $Df_p \oplus Df_p$ must preserve orientation. Thus $I_p(g) = 1$, and the same argument shows $I_p(g^m) = 1$ if $p \in \text{Fix}(g^m) \cap \Omega_i$. Consequently $L(\bar{g}^m) = \Sigma I_p(g^m) = N_m$, the number of fixed points of f^m in Ω_i. Therefore the zeta function of $f|\Omega_i$, $\zeta_i(f)$, equals the rational function $Z(\bar{g})$.   Q.E.D.

The following result gives expressions for $\zeta(f)$ and $Z(f)$ as infinite products when f has only hyperbolic periodic points. If in addition f has only finitely many periodic points the products are finite and give expressions for $\zeta(f)$ and $Z(f)$ as rational functions.

(5.19) PROPOSITION. *If $f \colon M \longrightarrow M$ is a diffeomorphism with all periodic points hyperbolic, then as formal power series in t,*

(a) $\zeta(f) = \Pi_\gamma (1 - t^{p(\gamma)})^{-1}$ *where the product is taken over all periodic orbits γ and $p(\gamma)$ is the (least) period of γ.*

(b) $Z(f) = \Pi_\gamma (1 - \Delta_\gamma t^{p(\gamma)})^{(-1)^{u(\gamma)+1}}$ *where Δ_γ is 1 if $Df_x^{p(\gamma)} \colon E_x^u \longrightarrow E_x^u$ preserves orientation for $x \in \gamma$ and -1 otherwise, and $u(\gamma) = \dim E_x$, $x \in \gamma$.*

If f has hyperbolic chain recurrent set and Ω_i is a basic set then the same formulas are valid for ζ_i and Z_i if the products are taken over all $\gamma \subset \Omega_i$.

PROOF. Since all periodic points are hyperbolic and M is compact, it follows that the points of period $\leqslant n$ are isolated and hence a finite set for any fixed n. If γ is a single orbit of period p then it is easy to check that $\zeta(f|\gamma) = (1 - t^p)^{-1}$ (since for example $f|\gamma$ is conjugate to a subshift of finite type corresponding to a permutation matrix and we can apply (5.3)). We now fix n and let $\{\gamma_1, \ldots, \gamma_s\}$ be the set of periodic orbits with period $p(\gamma_i) \leqslant n$ and let $K = \bigcup_{i=1}^s \gamma_i$; then $\zeta(f|K) = \Pi_{i=1}^s (1 - t^{p(\gamma_i)})^{-1}$.

But $N_n(f)$ equals $N_n(f|K)$, since any fixed point of f^n is in K. Thus the coefficient of t^n in $\zeta(f)$ is the same as the coefficient of t^n in $\zeta(f|K) = \Pi(1 - t^{p(\gamma_i)})^{-1}$. However, since $(1 - t^p)^{-1} = 1 + t^p + t^{2p} + \cdots$, the coefficient of t^n in $\Pi_{i=1}^s (1 - t^{p(\gamma_i)})^{-1}$ is the same as the coefficient of t^n in $\Pi_\gamma (1 - t^{p(\gamma)})^{-1}$, where the product is taken over all periodic orbits γ. Thus we have shown that the coefficients of t^n in $\zeta(f)$ and $\Pi_\gamma (1 - t^{p(\gamma)})^{-1}$ are the same. This proves (a).

The proof of (b) is similar; we use (5.7), which says that if γ has period p and $x \in \gamma$ then the fixed point index is $(-1)^{u(\gamma)}\Delta_\gamma$ where $\Delta_\gamma = +1$ if $Df^p: E_x^u \longrightarrow E_x^u$ preserves orientation and $\Delta_\gamma = -1$ if orientation is reversed. Now $Z(f) = \exp(\Sigma_{m=1}^\infty (1/m)L(f^m)t^m)$ and $L(f^m) = \Sigma I_x(f^m)$ where this sum is over all $x \in \mathrm{Fix}(f^m)$. Let K be as above and define $\rho = \exp(\Sigma(1/m)L_m(K)t^m)$ where $L_m(K)$ is the sum of $I_x(f^m)$ for all $x \in \mathrm{Fix}(f^m) \cap K$. Then for $m \leqslant n$ we have $L_m(K) = L(f^m)$ since all points of period $\leqslant n$ are in K. Thus the coefficient of t^n in ρ is the same as the coefficient of t^n in $Z(f)$.

But $L_m(K) = \Sigma_{i=1}^s L_m(\gamma_i)$ where $L_m(\gamma_i) = \Sigma I_x(f^m)$; the sum being taken over $x \in \gamma_i \cap \mathrm{Fix}(f^m)$. Hence

$$\rho = \prod_{i=1}^s \exp\left(\sum_{m=1}^\infty \frac{1}{m} L_m(\gamma_i) t^m \right).$$

Since

$$L_m(\gamma) = \begin{cases} 0 & \text{if } m \not\equiv 0 \bmod p(\gamma) \\ p(\gamma)(-1)^{u(\gamma)}\Delta^{m/p(\gamma)} & \text{if } m \equiv 0 \bmod p(\gamma), \end{cases}$$

one checks easily that $\rho = \Pi_{i=1}^s (1 - \Delta_{\gamma_i} t^{p(\gamma_i)})^{(-1)^{u(i)+1}}$. Thus, as before, the coefficient of t^n in ρ is also the same as the coefficient of t^n in $\Pi_\gamma (1 - \Delta_\gamma t^{p(\gamma)})^{(-1)^{u(\gamma)+1}}$, and (b) is proved. The proof for f restricted to a single basic set is similar. Q.E.D.

This result to a certain extent relates $\zeta(f)$ and $Z(f)$, but we want to pursue the connection between these two functions further since $\zeta(f)$ is an invariant of the dynamics while $Z(f)$ is a homological invariant. Of course, relating the dynamics of f to homological invariants is our long range goal. In (5.17) we established a close relationship between $\zeta_i(f)$ and $Z_i(f)$ if $Df: E^u(\Omega_i) \longrightarrow E^u(\Omega_i)$ preserved or reversed an orientation of E^u. This however is a very restrictive hypothesis.

By weakening the invariant $Z_i(t)$ somewhat we can make it an invariant of the dynamics of $f|\Omega_i$ and the index $u(i)$ of Ω_i.

(5.20) DEFINITION. If $f: M \longrightarrow M$ has hyperbolic chain recurrent set then the *reduced homology zeta function* $z(f)$ is defined to be the rational function of t with coefficients in $Z/2Z$ obtained by reducing all coefficients of $Z(f)$ mod 2. This makes sense since $Z(f)$ is a quotient of integer polynomials whose constant terms are 1. If Ω_i is a basic set we analogously define $z_i(f)$ to be the mod 2 reduction of $Z_i(f)$.

(5.21) THEOREM [F2]. *If $f: M \longrightarrow M$ has hyperbolic chain recurrent set and Ω_i is a basic set with index u then the following are equal:*

(a) *The rational function $\zeta_i(f)^{(-1)^u}$ with all coefficients reduced* mod 2,

(b) $z_i(f)$,

(c) *the function $\Pi_{k=0}^n \det(I - f_{*k}t)^{(-1)^{k+1}}$ where $f_{*k}: H_k(M_i, M_{i-1}; F_2) \longrightarrow H_k(M_i, M_{i-1}; F_2)$ is induced by f, $\{M_i\}$ is a filtration associated to f, and $F_2 = Z/2Z$.*

PROOF. To show that (a) equals (b) we note by (5.19)

$$Z_i(f)^{(-1)^u} = \prod(1 - \Delta_\gamma t^{p(\gamma)})^{-1} \quad \text{and} \quad \zeta_i(f) = \prod(1 - t^{p(\gamma)})^{-1},$$

where both products are taken over all periodic orbits in Ω_i. Clearly these should be the same when reduced mod 2 if we can make sense of the infinite products. We do this by considering formal power series.

Let $Z[t]$ be the ring of integer polynomials and let S be the multiplicative set $1 + tZ[t]$. Then $S^{-1}Z[t]$ will denote the ring of fractions of $Z[t]$ by S. Since for the inclusion $Z[t] \to Z[[t]]$ into formal power series the image of each element of S is invertible there is a unique extension of the inclusion to a homomorphism $\alpha: S^{-1}Z[t] \to Z[[t]]$ (see [L, pp. 66–69] for this).

Similarly if $F_2 = Z/2Z$ we have $F_2[t]$, and its localization at (t), $F_2[t]_{(t)}$, which is just $F_2[t]$ with $1 + tF_2[t]$ inverted. There is an extension of the inclusion $F_2[t] \to F_2[[t]]$ to $\beta: F_2[t]_{(t)} \to F_2[[t]]$. The homomorphism β is injective since it is injective on polynomials.

Let $\phi: Z[t] \to F_2[t]$, $\psi: S^{-1}Z[t] \to F_2[t]_{(t)}$ and $\theta: Z[[t]] \to F_2[[t]]$ all denote reduction of coefficients mod 2. Then we have the following commutative diagram of homomorphisms

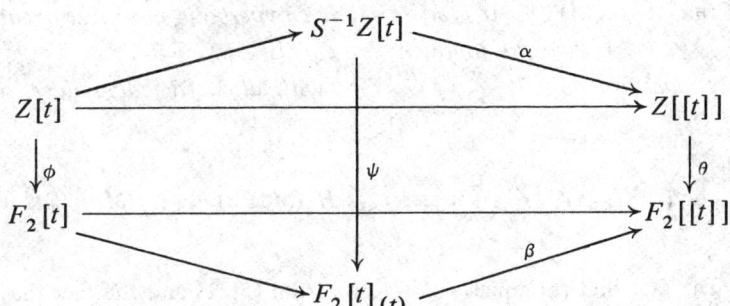

where the unlabelled arrows are the natural inclusions. The diagram is commutative because it commutes for polynomials. By (5.16) and (5.19) the rational functions ζ_i and Z_i are in $S^{-1}Z[t]$. The assertion of our theorem is that $\psi(Z_i) = \psi(\zeta_i^{(-1)^u})$. Considering the diagram and the fact that β is injective, it suffices to show that $\theta \circ \alpha(Z_i^{(-1)^u}) = \theta \circ \alpha(\zeta_i)$.

To do this we show they have the same coefficient of t^n. Let $[\gamma_1, \ldots, \gamma_s]$ be the set of periodic orbits in Ω_i with $p(\gamma_i) \leqslant n$. Then the coefficient of t^n in $\alpha(Z_i^{(-1)^u})$ is the same as that in $\alpha(\rho)$ where $\rho = \prod_{i=1}^s (1 - \Delta_{\gamma_i} t^{p(\gamma_i)})^{-1}$. Likewise the coefficient of t^n in $\alpha(\zeta_i)$ is the same as that in $\alpha(\hat{\rho})$ where $\hat{\rho} = \prod_{i=1}^s (1 - t^{p(\gamma_i)})^{-1}$. But $\psi(\rho) = \psi(\hat{\rho})$, so $\theta \circ \alpha(\rho) = \beta \circ \psi(\rho) = \beta \circ \psi(\hat{\rho}) = \theta \circ \alpha(\hat{\rho})$ and it follows that the coefficient of t^n in $\theta \circ \alpha(Z_i^{(-1)^u})$ is equal to the coefficient of t^n in $\theta \circ \alpha(\zeta_i)$. This proves that (a) equals (b).

To prove that (b) equals (c) we note that if $e: C \to C$ is an endomorphism of a finitely generated free abelian group then the function α defined by $\alpha(e) = \det(I - et)$ has values in the group of nonzero rational functions of t and is additive on short exact sequences, so we can apply (5.9). To do this we let $C = \{C_i\}$ be a finitely generated free chain complex whose homology is $H_*(M_i, M_{i-1})$ and let $\{e_k: C_k \rightarrow C_k\}$ be a chain map inducing f_{*k}: $H_k(M_i, M_{i-1}) \rightarrow H_k(M_i, M_{i-1})$ on homology. Then

$$Z_i(f) = \prod_{k=0}^{n} \det(I - f_{*k}t)^{(-1)^k} \quad \text{by (5.16)}$$

$$= \prod_{k=0}^{n} \det(I - e_k t)^{(-1)^k} \quad \text{by (5.9).}$$

Now if $\hat{e}_k: C_k \otimes F_2 \rightarrow C_k \otimes F_2$ is e_k reduced mod 2 then another application of (5.9) says

$$\prod_{k=0}^{n} \det(I - \hat{e}_k t)^{(-1)^{k+1}} = \prod_{k=0}^{n} \det(I - \hat{f}_{*k} t)^{(-1)^{k+1}}$$

where $\hat{f}_{*k}: H_k(M_i, M_{i-1}; F_2)$ is induced by f. Since $\det(I - \hat{e}_k t)$ clearly is the mod 2 reduction of $\det(I - e_k t)$, we have $z_i(f) = \prod \det(I - \hat{e}_k t)^{k+1} = \prod \det(I - \hat{f}_{*k} t)^{(-1)^{k+1}}$.   Q.E.D.

We have the following corollary relating $z(f)$ to $\zeta(f)$.

(5.22) COROLLARY [F2]. *If $f: M \rightarrow M$ has hyperbolic chain recurrent set and basic sets $\{\Omega_i\}$, $1 \leqslant i \leqslant l$, then the following are equal:*

(a) *The rational function $\prod_{i=1}^{l} \zeta_i(f)^{(-1)^{u(i)}}$ with all coefficients reduced* mod 2, *where* $u(i) = $ *index of Ω_i,*

(b) $z(f)$,

(c) $\prod_{k=0}^{n} \det(I - f_{*k}t)^{(-1)^{k+1}}$ *where $f_{*k}: H_k(M; F_2) \rightarrow H_k(M; F_2)$ is induced by f and $F_2 = Z/2Z$.*

PROOF. The fact that (a) equals (b) follows from (5.21) and the fact that $z(f) = \prod_i z_i(f)$ (since $Z(f) = \prod_i Z_i(f)$ by (5.17)).

To show (b) equals (c) one uses the same argument as in (5.21), with C a chain complex for M.   Q.E.D.

This result has applications to what one might call the bifurcation problem for basic sets. We know that $z(f) = \prod z_i(f)$ and $z(f)$ is a homotopy invariant of f while $z_i(f)$ depends only on $f|\Omega_i$ and the index $u(i)$ of Ω_i. It follows that if f and g are homotopic then $\prod z_i(f) = \prod z_i(g)$.

Several special cases of this give partial answers to interesting questions:

(1) When can an isotopy remove a basic set Ω_i of f while leaving all others unchanged? A necessary condition is that $z_i(f) = 1$.

(2) When can an isotopy "cancel" two basic sets Ω_i and Ω_j leaving all others unchanged? A necessary condition is $z_i(f)z_j(f) = 1$.

(3) When can an isotopy of f to g change a basic set $f: \Omega_i \rightarrow \Omega_i$ to a different basic set $g: \Omega_i' \rightarrow \Omega_i'$ leaving others unaltered? A necessary condition is $z_i(f) = z_i(g)$.

More detailed applications will be given in Chapter 6. For the moment we consider only the following example.

(5.23) EXAMPLE. In [Wm1] R. F. Williams showed that any subshift of finite type $\sigma(A)$: $\Sigma_A \rightarrow \Sigma_A$ with A irreducible can be realized as a basic set of a diffeomorphism of S^3. We give an example of a shift which cannot be realized as a basic set of a diffeomorphism of any S^n in such a way that all other basic sets are finite (i.e., periodic orbits), and which cannot occur at all as a basic set of a diffeomorphism of a surface with hyperbolic chain recurrent set.

If

$$A = \begin{pmatrix} 0 & 1 & 0 \\ 0 & 0 & 1 \\ 1 & 1 & 0 \end{pmatrix}$$

then A is an irreducible matrix and $\det(I - At) = 1 - t^2 - t^3$. Thus if $f|\Omega_i$ is conjugate to $\sigma(A)$, $z_i(f) = (1 + t^2 + t^3)$ or its reciprocal.

It is easy to check that $(1 + t^2 + t^3)$ is an irreducible factor of $1 + t^7$ in $F_2[t]$ and hence in the algebraic closure of F_2 its roots are three of the seven seventh roots of unity. On the other hand if Ω_j is a basic set which is a periodic orbit of period p then $z_j(f) = (1 + t^p)^{\pm 1}$, the exponent depending on $u(j) = $ index of Ω_j. In the algebraic closure of F_2, $(1 + t^p)$ has either no seventh roots of unity or all seven of them. Hence, since $z(f) = (1 + t)^{-2}$ or 1 for any diffeomorphism f: $S^n \rightarrow S^n$, it is impossible to have $z_i(f)\Pi_j z_j(f) = z(f)$ if all basic sets are finite except Ω_i.

The fact that σ: $\Sigma_A \rightarrow \Sigma_A$ cannot be conjugate to f restricted to a basic set for f: $M^2 \rightarrow M^2$ is a consequence of the following discussion. We want to investigate what kinds of reduced zeta functions can occur for diffeomorphisms of surfaces.

(5.24) THEOREM [Bl-F2]. *Suppose that Λ is a basic set of a surface diffeomorphism f: $M^2 \rightarrow M^2$ which has hyperbolic chain recurrent set. Then there exists a diffeomorphism g: $N \rightarrow N$ of a closed surface such that*

(a) *The diffeomorphism g has hyperbolic chain recurrent set.*

(b) *The only basic set of index 1 of g is Λ' and $f|\Lambda$ is topologically conjugate to $g|\Lambda'$.*

(c) *All other basic sets of g have index 0 or 2, and hence are periodic sources or sinks.*

PROOF. Choose a filtration $\{M_i\}$ of M^2 associated to f and suppose $\Lambda \subset M_i - M_{i-1}$. We first eliminate any index one basic sets "above" Λ in the filtration, i.e., in $M - M_i$. Let $X = \text{cl}(M_i - f(M_i))$. The Euler characteristic

$$\chi(X) = \chi(M_i) - \chi(f(M_i)) = 0.$$

If some components of X have positive Euler characteristic and some negative then some component of X is a disk D since no other surface with boundary has positive Euler characteristic. In this case add a disk D_0 to M_i, attaching it to $f^{-1}(\partial D) \subset \partial M_i$ and extend f to $M_i \cup D_0$ by making $f(D_0) = D$. Repeating this we obtain an embedding f': $M_i' \rightarrow M_i'$ such

that $X' = M_i' - f'(M_i)$ has no component a disk and hence each component has Euler characteristic 0. Every component of X' must have at least two boundary components (one in
∂M_i and one in $\partial f(M_i)$) so no component is a Moebius strip. Thus $X' = \bigcup A_i$ where each
A_i is an annulus with one boundary circle C_j^0 in ∂M_i and one C_j^1 in $\partial f(M_i)$. Form N_0 by
attaching a disk D_j to each C_j^0 and extend f' to $f_0: N_0 \rightarrow N_0$ by mapping D_j to $D_k \cup A_k$
where k is chosen so $f(C_j^0) = C_k^1$. We can do this with a periodic source at the center of
each D_j. It is easy to see that in a filtration for $f_0: N_0 \rightarrow N_0$ the only basic sets above N_0
are periodic sources.

   If we now repeat this process on the diffeomorphism $f_0^{-1}: N_0 \rightarrow N_0$ we arrive at the
desired $g: N \rightarrow N$.   Q.E.D.

   In [Bl-F2] this result and the reduced zeta function were used to show that certain
subshifts (including the one in (5.23) above) could not occur as basic sets of surface diffeomorphisms. These results were significantly improved by D. Fried who characterized reduced
zeta functions for surface diffeomorphisms. Note that the subshift in example (5.23) above
would have to have $z_i(f) = 1 + t^2 + t^3$ which does not satisfy the criterion below and
hence this subshift cannot be a basic set of a surface diffeomorphism.

   (5.25) THEOREM [Fr3]. *Suppose* $f: M^2 \rightarrow M^2$ *is a surface diffeomorphism with
hyperbolic chain recurrent set and* Ω_i *is a zero dimensional basic set. If* $\underset{\sim}{p}(t) = z_i(f)$ *is its
reduced zeta function then*
   (a) $p(0) = 1 \in F_2$, *and*
   (b) $p(t^{-1}) = t^k p(t)$ *for some* $k \in Z$.
Conversely given any polynomial $p(t) \in F_2[t]$ *satisfying* (a) *and* (b), *there is a Smale diffeomorphism* g *of a closed surface with* $p(t)$ *the reduced zeta function of a basic set.*

   PROOF. We first consider necessity of (a) and (b). Property (a) is immediate from
(5.21). To show (b) we construct a diffeomorphism $g: N \rightarrow N$ as in (5.24) with a single
basic set Λ of index one and topologically conjugate to $f|\Omega_i$. Now by (5.22)

$$\frac{p(t)}{\prod(1 \pm t^{q_j})} = \frac{\det(I - g_{*1} t)}{\det(I - g_{*0} t)\det(I - g_{*2} t)}$$

where $g_{*i}: H_i(N; F_2) \rightarrow H_i(N; F_2)$ is induced by g. Let $h(t) = \det(I - g_{*1} t)$, and note g_{*1} preserves the nonsingular intersection pairing $\langle\,,\rangle: H_1(N; F_2) \times H_1(N; F_2) \rightarrow F_2$. If J is a
matrix for this pairing and A a matrix for g_{*1} then $A^t JA = J$ so $A^t = JA^{-1}J^{-1}$. Hence

$$h(t^{-1}) = \det(I - At^{-1}) = \det(A^{-1}t)\det(A^{-1}t - I) = t^k h(t).$$

Since $\det(I - g_{*0} t)$ and $\det(I - g_{*2} t)$ both have the form $\Pi(1 + t^{p_i})$, it follows that $p(t^{-1})$
$= t^r p(t)$ for some $r \in Z$.

   We sketch the proof of the converse, supposing first that $p(t)$ has even degree. Then
choose a diffeomorphism $f: M^2 \rightarrow M^2$ such that

$$p(t) = \det(I - f_{*1} t)$$

where $f_{*1}: H_1(M^2; F_2) \longrightarrow H_1(M^2; F_2)$. (The proof that it is always possible to choose such an f is left to the reader.)

According to Lemma (B.1) of Appendix B we can isotope f to a fitted Smale diffeomorphism with a single fixed sink, a single fixed source and basic sets of index 1. Now applying arguments like the proof of (4.7) (which is (B.12) in Appendix B) we can push loops in one-handles through other one-handles making even increases in the geometric intersection matrix until it is strictly positive. Then there is only one basic set of index one. From (5.22) its reduced zeta function $z_1(f)$ satisfies

$$\frac{z_1(f)}{(1+t)^2} = \frac{p(t)}{\det(I - f_{*0}t)\det(I - f_{*1}t)}$$

so $z_1(f) = p(t)$ as desired.

In case the original $p(t)$ has odd degree then $p(t) = (1 + t)q(t)$ for some $q(t) \in F_2[t]$ (this is because (b) implies $p(t)$ is symmetric, i.e., $p(t) = 1 + a_1 t + a_2 t^2 + \cdots + a_n t^n + a_n t^{n+1} + \cdots + a_1 t^{2n} + a_0 t^{2n+1}$, so $p(1) = 0$). Now we apply the same argument as above using $q(t)$ and including 1 sink and 2 sources, to obtain the desired result.   Q.E.D.

Chapter 6. Morse Inequalities

In Chapter 2 we discussed the classical Morse inequalities for the gradient flow of a Morse function. They relate homological invariants of M to the indices of the rest points of the flow.

The algebraic basis of these inequalities is the fact that if $b_0, b_1, \ldots, b_n$ are the ranks of the homology of a finite rank chain complex and $c_0, c_1, \ldots, c_n$ are the ranks of the chain complex, then

$$c_q - c_{q-1} + \cdots \pm c_0 \geqslant b_q - b_{q-1} + \cdots \pm b_0$$

must hold for all q. In fact these are necessary and sufficient conditions that there exist a chain complex $C = \{C_i\}$ with rank $C_i = c_i$ and rank $H_i(C) = b_i$ (see (2.13)).

If the chain complex is provided with a chain map τ and λ is an eigenvalue of τ we can define

$c_i(\lambda)$ = dimension of the generalized eigenspace for λ in C_i, and

$b_i(\lambda)$ = dimension of the generalized eigenspace for λ in $H_i(C)$ (with the map τ_*).

It is then easy to see that the Morse inequalities above also hold for $\{c_i(\lambda)\}$ and $\{b_i(\lambda)\}$. But since $c_i(\lambda)$ is the exponent of $(t - \lambda)$ in the characteristic polynomial of $\tau|C_i$ and $b_i(\lambda)$ is the exponent of $(t - \lambda)$ in the characteristic polynomial of $\tau_*|H_i(C)$ one can consolidate these inequalities (for different λ's) by considering them as inequalities on the exponents of linear factors of rational functions from the characteristic polynomials of τ_i and τ_{*i}.

In view of the results of Chapter 5 it should be clear that for our purposes the polynomials $\det(I - \tau_i t)$ and $\det(I - \tau_{*i} t)$ are more useful than the characteristic polynomials. Since, for any matrix A, with characteristic polynomial $h(t)$, $\det(I - At) = t^k h(1/t)$ for some $k > 0$, it is clear that if $\lambda \neq 0$ then $c_i(\lambda)$ is the exponent of $(1 - \lambda t)$ in $\det(I - \tau_i t)$ and $b_i(\lambda)$ is the exponent of $(1 - \lambda t)$ in $\det(I - \tau_{*i} t)$.

Consolidating the inequalities on these exponents we can say that for each q there exists an undetermined polynomial $P(t)$ such that

$$P(t)^{(-1)^{q+1}} \prod_{k=0}^{q} \det(I - \tau_k t)^{(-1)^k} = \prod_{k=0}^{q} \det(I - \tau_{*k} t)^{(-1)^k}.$$

Thus the alternating sums have become alternating products of polynomials and the "inequality" is expressed in the fact that $P(t)$ is a polynomial, i.e., all its factors have positive exponents.

This is the algebraic idea for the results of this chapter. However, we are interested in relating the maps $f_*: H_*(M_i, M_{i-1}) \longrightarrow H_*(M_i, M_{i-1})$ to the maps $f_*: H_*(M) \longrightarrow H_*(M)$ and in the generality we consider, $H_*(M_i, M_{i-1})$ may well not be a chain complex with homology $H_*(M)$. Although we will derive Morse inequalities by a different approach, the idea of these heuristic remarks can be carried out as a complete proof utilizing the spectral sequence associated to a filtration $\{M_i\}$ in place of the chain complex. For the reader familiar with spectral sequences this might give a simpler proof of (6.5).

We begin with a lemma about the dimension of $W^u(\Omega_i)$ for a basic set Ω_i.

(6.1) LEMMA [F3]. *If $f: M \longrightarrow M$ has a hyperbolic chain recurrent set and Ω_i is a basic set of index $u(i)$, then*

$$\dim \Omega_i + u(i) \geqslant \dim W^u(\Omega_i)$$

A proof of this can be found in [F3] where it is Lemma 5. We now prove a generalization of (4.11(b)). It is a slight improvement of results of Bowen [B3].

(6.2) LEMMA [F3]. *Suppose $f: M \longrightarrow M$ has a hyperbolic chain recurrent set, with basic sets $\{\Omega_i\}$ and associated filtration $\{M_i\}$.*

(a) *If $\dim W^u(\Omega_i) < k$ (in particular if $\dim \Omega_i + u(i) < k$), then $f_{*k}: H_k(M_i, M_{i-1}; F) \longrightarrow H_k(M_i, M_{i-1}; F)$ is nilpotent for any coefficient field F.*

(b) *Dually, if $\dim W^s(\Omega_i) < (\dim M) - k$, or if $u(i) - \dim \Omega_i > k$, then $f_{*k}: H_k(M_i, M_{i-1}; F) \longrightarrow H_k(M_i, M_{i-1}; F)$ is nilpotent.*

PROOF. (a) If f_{*k} is not nilpotent then it has a nonzero eigenvalue λ, and hence $f_k^*: H^k(M_i, M_{i-1}; F) \longrightarrow H^k(M_i, M_{i-1}; F)$ has λ as an eigenvalue. Let $X = \bigcap_{n \geqslant 0} f^n(M_i)$ and $A = \bigcap_{n \geqslant 0} f^n(M_{i-1})$. Since the Čech cohomology $\check{H}^k(X, A)$ is $\varinjlim_n H^k(f^n(M_i), f^n(M_{i-1}))$ (see (VIII. 6) of [D1]) it is easy to see that $f_k^*: \check{H}^k(X, A; F) \longrightarrow \check{H}^k(X, A; F)$ has λ as an eigenvalue also. However $X - A = W^u(\Omega_i)$ has dimension $< k$ and it follows using excision that $\check{H}^k(X, A_0; F) = 0$ for any closed neighborhood A_0 of A (see [H-W, p. 152]). This implies (by continuity of Čech cohomology) that $\check{H}^k(X, A; F) = 0$ which contradicts the assumption that λ was an eigenvalue. Hence if $\dim W^u(\Omega_i) < k$, $f_{*k}: H_k(M_i, M_{i-1}; F) \longrightarrow H_k(M_i, M_{i-1}; F)$ is nilpotent. To complete (a) we note (6.1) says $\dim \Omega_i + u(i) > \dim W^u(\Omega_i)$ so we have the same conclusion if $\dim \Omega_i + u(i) < k$.

To prove (b) we consider $g = f^{-1}: M \longrightarrow M$ with associated filtration $\{\widetilde{M}_i\}$ where $\widetilde{M}_i = \mathrm{cl}(M - M_i)$. From the proof of (a) we conclude that if $n = \dim M$ and $\dim W^u(\Omega_i; g)$ (which equals $\dim W^s(\Omega_i; f)$) is less than $n - k$ then

$$g^*: H^{n-k}(\widetilde{M}_{i-1}, \widetilde{M}_i; F) \longrightarrow H^{n-k}(\widetilde{M}_{i-1}, \widetilde{M}_i; F)$$

is nilpotent. If M is orientable over F then by Poincaré duality g^* is similar to $\pm f_{*k}: H_k(M_i, M_{i-1}; F) \longrightarrow H_k(M_i, M_{i-1}; F)$ so f_{*k} is nilpotent. If M is not orientable one must use a double cover and the homology transfer to prove the same result. We omit the details.

To complete (b) we let $s(i)$ = fiber dimension of $E^s(\Omega_i; f)$. Thus applying (6.1) to g we get $s(i) + \dim \Omega_i \geqslant \dim W^u(\Omega_i; g) = \dim W^s(\Omega_i; f)$. Hence

$$u(i) - \dim \Omega_i > k \quad \text{implies} \quad n - s(i) - \dim \Omega_i > k$$

so $n - k > s(i) + \dim \Omega_i \geqslant \dim W^s(\Omega_i; f)$. It follows that $u(i) - \dim \Omega_i > k$ also implies f_{*k} is nilpotent. Q.E.D.

In order to prove Morse inequalities we will need a hypothesis on basic sets which says they can be divided into two groups, those that contribute only to homology in dimensions less than or equal to some number q and those that contribute to homology only in dimensions greater than q.

(6.3) DEFINITION. If $f: M \longrightarrow M$ has a hyperbolic chain recurrent set we will say that the basic sets are *homologically split at* q over the field F if for any filtration $\{M_j\}$ associated to f and any basic set Ω_i, the maps $f_{*k}: H_k(M_i, M_{i-1}; F) \longrightarrow H_k(M_i, M_{i-1}; F)$ are nilpotent for all $k > q$ if $u(i) \leqslant q$, and nilpotent for all $k \leqslant q$ if $u(i) > q$.

Lemma (6.2) gives conditions which will guarantee that basic sets are homologically split at q. For example, if for an integer q it is true that each basic set Ω_i with index $u(i) \leqslant q$ satisfies $\dim W^u(\Omega_i) \leqslant q$ and each basic set Ω_j with $u(j) > q$ satisfies $\dim W^s(\Omega_j)$ $< n - q$, where $n = \dim M$, then the basic sets are homologically split at q. We also have the following.

(6.4) PROPOSITION. *If all basic sets of* $f: M \longrightarrow M$ *have dimension zero then they are homologically split at* q *for all* q *and any coefficients.*

PROOF. From (6.2) it follows that $f_{*k}: H_k(M_i, M_{i-1}; F) \longrightarrow H_k(M_i, M_{i-1}; F)$ is nilpotent unless $k = u(i)$. Hence for any q, f_{*k} is nilpotent for $k \leqslant q$ if $u(i) > q$ and nilpotent for $k > q$ if $u(i) \leqslant q$. Q.E.D.

We can now prove the promised Morse inequalities for the zeta functions $Z_i(f)$.

The results of Chapter 5 related the functions $Z_i(f)$ (and to some extent $\zeta_i(f)$) to the induced map $f_*: H_*(M_i, M_{i-1}) \longrightarrow H_*(M_i, M_{i-1})$. It is much more valuable, however, to establish relationships with $f_*: H_*(M) \longrightarrow H_*(M)$ since it is not always easy to determine the filtration manifolds or their homologies.

(6.5) THEOREM [F2] (MORSE INEQUALITIES). *If* $f: M \longrightarrow M$ *has a hyperbolic chain recurrent set and its basic sets are homologically split at* q *over R, then there exists an integer polynomial* $P(t)$ *such that*

$$P(t)^{(-1)^q} \prod_{u(i) \leqslant q} Z_i(f) = \prod_{k=0}^{q} \det(I - f_{*k} t)^{(-1)^{k+1}}$$

where $f_{*k}: H_k(M; R) \longrightarrow H_k(M; R)$ *is induced by* f.

PROOF. Suppose $M = M_l \supset \cdots \supset M_1 \supset M_0 = \varnothing$ is a filtration for f. Define

$$\eta^q(M_i, M_j) = \prod_{0 \leqslant k \leqslant q} \det(I - f_* t)^{(-1)^{k+1}},$$

where $f_{*k}: H_k(M_i, M_j; R) \longrightarrow H_k(M_i, M_j; R)$ is the map induced by f. Consider now the exact sequence $0 \longrightarrow B \longrightarrow H_q(M_j) \longrightarrow H_q(M_i) \longrightarrow H_q(M_i, M_j) \longrightarrow H_{q-1}(M_j) \longrightarrow \cdots$, where $B = \ker(i_*: H_q(M_j) \longrightarrow H_q(M_i))$ and the remainder of the sequence is the exact sequence of the pair (M_i, M_j). Note $f_{*q}(B) \subset B$ and let $P_{i,j} = \det[I - (f_{*q}|B)t]$. We consider this exact sequence as a chain complex with trivial homology and with chain map the endomorphisms induced by f. We then apply (5.9) with the map α defined on an endomorphism e by $\alpha(e) = \det(I - et)$ and taking values in the abelian group of nonzero rational functions of t under multiplication. It is easy to check using matrices that α is multiplicative on short exact sequences, so satisfies the hypothesis of (5.9). Applying it we obtain the result

$$P_{ij}^{(-1)^q} \cdot \eta^q(M_j) \cdot \eta^q(M_i)^{-1} \cdot \eta^q(M_i, M_j) = 1$$

Thus, if we set $j = i - 1$ and denote $P_{i,i-1}$ by P_i we have,

$$\eta^q(M_i, M_{i-1}) = P_i^{(-1)^{q+1}} \cdot \eta^q(M_i) \cdot \eta^q(M_{i-1})^{-1}.$$

Taking a product over $0 \leqslant i \leqslant l$ we get

$$\prod_{i=1}^{l} \eta^q(M_i, M_{i-1}) = \eta^q(M_l) \cdot \eta^q(M_0)^{-1} \prod_{i=1}^{l} P_i^{(-1)^{q+1}} = \eta^q(M) \cdot P^{(-1)^{q+1}}$$

where $P = \Pi_{i=1}^{l} P_i$, since $M_l = M$ and $M_0 = \varnothing$. Notice P is a polynomial with integer coefficients and constant term 1.

By hypothesis, if Ω_i is a basic set with $u(i) \leqslant q$ then $f_{*k}: H_k(M_i, M_{i-1}; R) \longrightarrow H_k(M_i, M_{i-1}; R)$ is nilpotent if $k > q$. That is, $\det(I - f_{*k}t) = 1$ if $k > q$. It follows that $\eta^q(M_i, M_{i-1}) = \Pi_{k=0}^{n} \det(I - f_{*k}t)^{(-1)^{k+1}}$ where $n = \dim M$ and $f_{*k}: H_k(M_i, M_{i-1}; R) \longrightarrow H_k(M_i, M_{i-1}; R)$. However, by (5.16(c)) this is equal to $Z_i(f)$, so $\eta^q(M_i, M_{i-1}) = Z_i(f)$ if $u(i) \leqslant q$.

On the other hand when $u(j) > q$, we have f_{*k} is nilpotent if $k \leqslant q$. So a similar argument shows $\eta^q(M_j, M_{j-1}) = 1$ if $u(j) > q$.

Thus

$$P^{(-1)^{q+1}} \eta^q(M) = \prod_{i=1}^{l} \eta^q(M_i, M_{i-1}) = \prod_{u(i) \leqslant q} Z_i(f).$$

Since by definition,

$$\eta^q(M) = \prod_{k=0}^{q} \det(I - f_{*k}t)^{(-1)^{k+1}},$$

where $f_{*k}: H_k(M; R) \longrightarrow H_k(M; R)$, we have the desired result.   Q.E.D.

This same proof can be carried out over the field $F_2 = Z/2Z$. In the equality of the theorem $P(t)$ is replaced by a polynomial with coefficients in F_2, $Z_i(f)$ is replaced by $z_i(f)$ and of course f_{*k} becomes the map induced on $H_k(M; F_2)$ by f. The proof is exactly the same using (5.21) in place of (5.16) and we obtain the following result.

(6.6) THEOREM [F2]. *Suppose $f: M \to M$ has a hyperbolic chain recurrent set and its basic sets are homologically split at q over F_2; then there exists a polynomial $p(t) \in F_2[t]$ such that*

$$p(t)^{(-1)^q} \prod_{u(i) \leqslant q} z_i(f) = \prod_{k=0}^{q} \det(I - f_{*k}t)^{(-1)^{k+1}}$$

where $f_{*k}: H_k(M; F_2) \to H_k(M; F_2)$ is induced by f.

The following result gives a necessary condition for a collection of "abstract" basic sets to be embedded as the basic sets of any diffeomorphism f of M (no matter what the homotopy class of f). By the degree of a rational function we mean the degree of the numerator minus the degree of the denominator. The following result was inspired by the remark of Smale [S1] that the degree of the homology zeta function is minus the Euler characteristic of M.

(6.7) PROPOSITION [F2]. *If $f: M \to M$ has a hyperbolic chain recurrent set with basic sets $\{\Omega_i\}_{i=1}^{l}$ then*

$$\sum_{i=1}^{l} \deg z_i(f) = -\chi(M)$$

where $\chi(M)$ is the Euler characteristic of M.

PROOF. From (5.22) we have

$$\prod_{i=1}^{l} z_i(f) = \prod_{k=0}^{n} \det(I - f_{*k}t)^{(-1)^{k+1}}$$

where $f_{*k}: H_k(M; F_2) \to H_k(M; F_2)$ is induced by f. The degree of the left-hand side of this equation is $\Sigma \deg z_i(f)$ and the degree of the right-hand side is

$$-\sum_{k=0}^{n} (-1)^k \deg(\det(I - f_{*k}t)).$$

Since f_{*k} is an isomorphism

$$\deg(\det(I - f_{*k}t)) = \dim H_k(M; F_2).$$

Thus

$$\sum_{i=1}^{l} \deg z_i(f) = -\sum_{k=0}^{n} (-1)^k \dim H_k(M; F_2) = -\chi(M). \quad \text{Q.E.D.}$$

The point of using the reduced zeta function $z_i(f)$ is that, unlike the homology zeta function $Z_i(f)$, it is an invariant of the topological conjugacy type of $f|\Omega_i$ together with the index $u(i)$. For example, we can ask if a certain isotopy class admits a Smale diffeomorphism whose basic sets are a given set of subshifts of finite type with prescribed indices. The functions $\{z_i(f)\}$ can be calculated from just this data, and then (6.6) gives necessary homological conditions that the isotopy class must satisfy.

Of course reducing mod 2 throws away a great deal of information, so it is perhaps unrealistic to expect these necessary conditions to be close to sufficient. Surprisingly however, it turns out that if we consider the homotopy class of the identity (which is the most important one) and only ask to specify the basic sets up to a finite power instead of exactly, the conditions of (6.6) are sufficient as well as necessary. Of course, for any Smale diffeomorphism the basic sets of index 0 and $n = \dim M$ must be finite and therefore be subshifts associated to permutation matrices.

(6.8) THEOREM [Ba1, F2]. *Suppose that an n dimensional manifold M admits a Morse function with the number of critical points of index k equal to $\dim H_k(M; F_2)$ for each k. Let $\sigma(B_i)$: $\Sigma_{B_i} \longrightarrow \Sigma_{B_i}$, $1 \leqslant i \leqslant l$, be subshifts of finite type with irreducible matrices, each of which is assigned an index $u(i)$, and suppose B_j is a permutation matrix if $u(j) = 0$ or n.*

Then there exists a Smale diffeomorphism $f: M \longrightarrow M$ homotopic to the identity, with basic sets $\{\Omega_i\}$, $1 \leqslant i \leqslant l$, of index $u(i)$ and with $f|\Omega_i$ topologically conjugate to $\sigma(B_i)^m$ for some $m > 0$ independent of i, if and only if the existence of such an f is compatible with the mod 2 Morse inequalities (6.6). More precisely, if $\beta_k = \dim H_k(M; F_2)$ and $d_k = \Sigma_{u(i)=k} \deg p_i(t)$ where $p_i(t) \in F_2[t]$ is the mod 2 reduction of $\det(I - B_i t)$, and the sum is over all i with $u(i) = k$, then necessary and sufficient conditions for the existence of f as described above are that

$$d_q - d_{q-1} + \cdots \pm d_0 \geqslant \beta_q - \beta_{q-1} + \cdots \pm \beta_0 \quad \text{for all } q.$$

This result was first proved in [F3] with the added assumption that $\dim M > 2$. The case $\dim M = 2$ was done by Steve Batterson in [Ba1]. Here, we give the proof of sufficiency only in the special case that $\dim M > 2$ and with the added assumption $B_1 = (1)$ and $u(1) = 0$, $B_l = (1)$ and $u(l) = n$, and $u(i) \neq 0$, n if $i \neq 1, l$. In other words we assume there is a single basic set of index 0 and of index n and both of them are fixed points.

PROOF. We first show sufficiency of the inequality. By hypothesis M admits a self-indexing Morse function $\varphi_0: M \longrightarrow R$ with β_k critical points of index k. Using (2.17) we form a new Morse function adding a number of pairs of cancelling critical points. In fact we add $\alpha_k = (d_k - d_{k-1} + \cdots \pm d_0) - (\beta_k - \beta_{k-1} + \cdots \pm \beta_0)$ pairs of index k and $k + 1$. The fact that $\alpha_k \geqslant 0$ follows from the inequality we are claiming is sufficient for the existence of f.

Clearly the new self-indexing Morse function $\varphi_1: M \longrightarrow R$ has $\beta_k + \alpha_k + \alpha_{k-1} = d_k$ critical points of index k. Let $f_1: M \longrightarrow M$ be a fitted diffeomorphism isotopic to the identity whose kth geometric intersection matrix is the $d_k \times d_k$ identity matrix, e.g., f_1 could be the time T map of a fitted flow, gradient-like with respect to φ_1 (see (B.3)).

We want to use (4.7) to change these geometric intersection matrices to $\{G(k)\}$ so that $\sigma(G(k))$ will be topologically conjugate to $\sigma(A_k)$ where A_k is the direct sum of the matrices $(B_i)^m$ with $u(i) = k$, and $m > 0$ is some fixed number independent of i. Before doing this we need two lemmas about congruences of integer matrices.

(6.9) LEMMA. *If $B^2 \equiv B$ (mod 4), then there exist nonnegative square integer matrices X, Y, D, with D diagonal, such that $XDY \equiv B$ (mod 4) and $DYX \equiv D$ (mod 4).*

PROOF. Let $\bar{B}$ be the mod 4 reduction of B, i.e., a matrix with entries in $Z/4Z$. If it is $n \times n$, we consider it as an endomorphism of M, the free module with n generators over $Z/4Z$. We let $R = \text{image}(\bar{B})$ and $K = \text{Ker}(\bar{B}) = \text{image}(I - \bar{B})$.

Clearly the map $x \mapsto (\bar{B}(x), (I - \bar{B})(x))$ is a direct sum decomposition of M into $R \oplus K$. By the fundamental theorem of Abelian groups R and K are the direct sum of copies of $Z/4Z$ and possibly $Z/2Z$, but because of the direct sum decomposition there can be no copies of $Z/2Z$. Thus R and K are free $Z/4Z$ modules of ranks, say r and k where $r + k = n$. If we form a basis for M with the first r elements a basis of R and the last k a basis of K, then clearly the matrix of the endomorphism with respect to this basis $\bar{D}$ will be diagonal (since it is the identity on R). If $\bar{X}$ is the change of basis matrix then we have $\overline{XDX}^{-1} = \bar{B}$. So if we let X, Y and D be nonnegative integral matrices which reduce mod 4 to $\bar{X}$, $\bar{X}^{-1}$, and $\bar{D}$ respectively then we have the desired matrices. Q.E.D.

(6.10) LEMMA. *If A is a direct sum of nonnegative irreducible matrices then there exist an integer $m > 0$ and (not necessarily square) nonnegative matrices R and S such that $A^m = RS$, and SR is congruent mod 2 to a diagonal matrix.*

PROOF. Since an irreducible matrix has a power which is a direct sum of strictly positive matrices (see [G, pp. 53–54]), we can assume by taking a power that A is a direct sum of positive matrices. By raising to a power again we can assume that $A^2 \equiv A$ (mod 4). This is because $A^n \equiv A$ (mod 4), $n > 2$, implies $(A^{n-1})(A^{n-1}) \equiv (A^{n-1}A)A^{n-2} \equiv AA^{n-2} \equiv A^{n-1}$. Now suppose $B_1, B_2, \ldots, B_m$ are the strictly positive direct summands of A; then $B_i^2 \equiv B_i$ (mod 4). By (6.9) there are square nonnegative matrices X_i, Y_i, D_i such that $X_i D_i Y_i \equiv B_i$ (mod 4) and $D_i Y_i X_i \equiv D_i$ (mod 4). Since $B_i^k \equiv B_i$ (mod 4), by choosing k sufficiently large all entries of B_i^k will be greater than those of $X_i D_i Y_i$ so we will have $B_i^k = X_i D_i Y_i + 4H_i$ where H_i is a nonnegative integral matrix.

We now note the matrix equations

$$(X_i, 2H_i)\begin{pmatrix} D_i Y_i \\ 2I \end{pmatrix} = X_i D_i Y_i + 4H_i = B_i^k$$

and

$$\begin{pmatrix} D_i Y_i \\ 2I \end{pmatrix}(X_i, 2H_i) = \begin{pmatrix} D_i Y_i X_i & 2D_i Y_i H_i \\ 2X_i & 4H_i \end{pmatrix}$$

which is congruent mod 2 to a diagonal matrix. Hence if we let

$$R_i = (X_i, 2H_i), \qquad S_i = \begin{pmatrix} D_i Y_i \\ 2I \end{pmatrix}$$

and let R be the direct sum of the R_i and S the direct sum of the S_i, these matrices will have the desired property.   Q.E.D.

We return now to the proof of (6.8). We notice that by (6.10) there is an $m(k) > 0$ such that $A_k^{m(k)}$ is strong shift equivalent to a matrix which is congruent mod 2 to a diagonal matrix. If $m = \Pi_k m(k)$ then A_k^m will also have this property. We choose $G(k)$ to be a matrix strong shift equivalent to A_k^m and congruent mod 2 to a diagonal matrix. Our aim is to construct a fitted diffeomorphism with geometric intersection matrices $\{G(k)\}$.

Suppose $G(k)$ is an $n_k \times n_k$ matrix. It is easy to see that the mod 2 reduction of $\det(I - G(k)t)$ equals the mod 2 reduction of $\det(I - A_k^m t)$ which has degree d_k. Thus, since the mod 2 reduction of $G(k)$ is diagonal, it has d_k ones on the diagonal and $n_k - d_k$ zeroes; in particular $n_k \geqslant d_k$.

We want now to construct a fitted diffeomorphism f_2 with geometric intersection matrices $\widetilde{G}(k)$ where $\widetilde{G}(k)$ is a diagonal matrix of zeroes and ones which has d_k ones on the diagonal and has size $\geqslant n_k \times n_k$. The construction of the desired diffeomorphism f then follows immediately by repeated application of (4.7). To produce f_2 we first use (2.17) again to add critical points until there are at least n_k handles of index k for each k. If $h_i(k)$ is one of the new handles added then $f_1^r(h_i(k)) \subset M_0$ for some $r > 0$, since $h_i(k)$ can be chosen disjoint from stable and unstable manifolds (for f_1) of dimension $< n$. Thus if we define $f_2 = f_1^r$, for some sufficiently large r, f_2 will have geometric intersection matrices $\{\widetilde{G}(k)\}$. Altering f_2 so that it is fitted (see (B.5) in Appendix B) gives the desired result.

This completes the proof that the inequality

$$d_q - d_{q-1} + \cdots \pm d_0 \geqslant \beta_q - \beta_{q-1} + \cdots \pm \beta_0$$

is sufficient for the existence of f. To prove that it is necessary we will show it is a consequence of the mod 2 Morse inequalities (6.6),

$$p(t)^{(-1)^q} \prod_{u(i) \leqslant q} z_i(f) = \prod_0^q \det(I - f_{*k} t)^{(-1)^{k+1}}.$$

We take the degree of each side of this equation (as we did in the special case (6.7)). Note that $(-1)^{k+1} d_k = \Sigma \deg z_i(f)$, the sum being over all i with $u(i) = k$. Also $\deg(\det(I - f_{*k})) = \beta_k$, so we have

$$(-1)^q \deg p(t) + \sum_{k=0}^q (-1)^{k+1} d_k = \sum_{k=0}^q (-1)^{k+1} \beta_k.$$

This gives the desired inequality since $\deg p(t) \geqslant 0$. Since the inequalities (6.6) imply this inequality we have also shown that compatibility with (6.6) is sufficient for the existence of f.   Q.E.D.

Chapter 7. Morse-Smale diffeomorphisms

In the two preceding chapters we developed necessary conditions for the existence of a diffeomorphism in a given homotopy class with certain prescribed behavior on its hyperbolic chain recurrent set R. In this chapter we show that in the case R is finite these conditions are often sufficient as well. We also investigate homological obstructions to the existence of f with a finite hyperbolic chain recurrent set R.

(7.1) DEFINITION. A diffeomorphism $f: M \longrightarrow M$ is called *Morse-Smale* provided its chain recurrent set R consists of a finite set of hyperbolic periodic points and f satisfies the transversality condition (see (1.9)).

In other words f is Morse-Smale provided it is a Smale diffeomorphism with finite R. The basic sets of a Morse-Smale diffeomorphism f are all hyperbolic periodic orbits. A basic set γ is completely described if we specify its index, its period p and whether $Df_*^p: E_x^u \longrightarrow E_x^u$ preserves or reverses orientation for some (and hence all) $x \in \gamma$.

(7.2) DEFINITION. If $f: M \longrightarrow M$ is a Morse-Smale diffeomorphism with periodic orbits $\gamma_1^k, \gamma_2^k, \ldots, \gamma_{r(k)}^k$ of index k, the *periodic data* of f is the array of integer pairs $\{(p_{ik}, \Delta_{ik})\}$, $1 \leqslant i \leqslant r(k)$, $0 \leqslant k \leqslant n$, where p_{ik} is the (least) period of the orbit γ_i^k and $\Delta_{ik} = \pm 1$ depending on whether $df^{p_{ik}}: E_x^u \longrightarrow E_x^u$, $x \in \gamma_i^k$, preserves or reverses orientation.

By calling it an array we indicate that an element of the periodic data is a pair of integers $(p, \pm 1)$ together with a position, e.g. the ikth, in the array. The integer k is called the *index* of the element (p_{ik}, Δ_{ik}).

We are interested in the existence of Morse-Smale diffeomorphisms with prescribed periodic data. Necessary conditions for this existence are given by the inequalities of (6.5).

(7.3) THEOREM [F2]. *If* $f: M \longrightarrow M$ *is a Morse-Smale diffeomorphism with periodic data* $\{(p_{ik}, \Delta_{ik})\}$ *then for each* q *there is a polynomial* $P_q(t)$ *such that*

$$P_q^{(-1)^q} \prod_{k \leqslant q} \det(I - f_{*k} t)^{(-1)^k} = \prod_{k \leqslant q} (I - \Delta_{ik} t^{p_{ik}})^{(-1)^k}$$

where $f_{*k}: H_k(M; R) \longrightarrow H_k(M; R)$ *is induced by* f.

PROOF. This is an immediate consequence of (6.5) and (5.19). Q.E.D.

For two dimensional manifolds these conditions are sometimes but not always sufficient. We have the following results of Batterson and Narasimhan.

(7.4) THEOREM [Ba2, N]. *Suppose that*

(a) $f: M^2 \longrightarrow M^2$ *is homotopic to the identity, where M^2 is a compact surface, or*

(b) $f: T^2 \longrightarrow T^2$ *is orientation preserving where T^2 is $S^1 \times S^1$.*

Then necessary and sufficient conditions for the existence of a Morse-Smale diffeomorphism with periodic data $\{(p_{ik}, \Delta_{ik})\}$ are that

(1) *There exist elements in the periodic data of index 0 and of index 2 and $\Delta_{i0} = \Delta_{i2} = 1$ for any i; and*

(2) $Z(f) = \Pi_{i,k}(1 - \Delta_{ik} t^{p_{ik}})^{(-1)^{k+1}}$.

The proof of this result with hypothesis (a) is due to C. Narasimhan [N] and with hypothesis (b) to S. Batterson [Ba2]. Batterson has also given necessary and sufficient conditions when $f: T^2 \longrightarrow T^2$ is orientation reversing [Ba3]. In this case, and in fact whenever a map $f: M^2 \longrightarrow M^2$ reverses orientation, the conditions of (7.3) will not be sufficient (see [Bl-F] and [H]).

We want now to apply our results on constructing fitted diffeomorphisms from Chapter 4 to the special case of Morse-Smale diffeomorphisms. For a fitted diffeomorphism to be Morse-Smale, the algebraic intersection matrix (and also the geometric intersection matrix) must have a special form.

(7.5) DEFINITION. A matrix will be called *virtual permutation* provided it has the form

$$\begin{pmatrix} P_1 & * & & & * \\ 0 & P_2 & * & & \\ & 0 & & & \\ & & & & * \\ 0 & & & 0 & P_r \end{pmatrix}$$

where each P_i has the form

$$\begin{pmatrix} 0 & 1 & 0 & & 0 \\ & 0 & 1 & 0 & \\ & & & & \\ 0 & & & & 1 \\ \pm 1 & 0 & & & 0 \end{pmatrix}.$$

(7.6) PROPOSITION. *A fitted diffeomorphism $f: M \longrightarrow M$ is Morse-Smale if each of its geometric intersection matrices $G(k)$ is a virtual permutation matrix. In this case the handles can be oriented so that each algebraic intersection matrix $A(k)$ will also be a virtual permutation matrix.*

PROOF. By (3.6) and (3.11), on each basic set f will be topologically conjugate to a subshift $\sigma(P)$: $\Sigma_P \longrightarrow \Sigma_P$ where P is a permutation matrix. Hence the chain recurrent set of f is finite.

To choose the orientations of handles so that $A(k)$ is a virtual permutation suppose that γ_i^k is the basic set corresponding to a permutation matrix P_i in $G(k)$, and γ_i^k is contained in the handles $h_j(k), h_{j+1}(k), \ldots, h_{j+p}(k)$, where $p = p_{ik}$ is the period of γ_i^k. We orient $h_j(k)$ arbitrarily (i.e., we orient $W_j^u(k)$) and then orient $h_{j+l}(k)$, $1 \leqslant l \leqslant p$, so that the orientations of $f^l(W_j^u(k))$ and $W_{j+l}^u(k)$ agree. The orientations of $f(W_{j+p}^u(k))$ and $W_j^u(k)$ agree or disagree depending on whether $\Delta_{ik} = \pm 1$. But the matrix $A(k)$ will have a block upper triangular form with the diagonal blocks being permutation matrices with perhaps a single minus sign in the lower left corner.   Q.E.D.

On the chain level, rather than the homology level, necessary conditions for the existence of a Morse-Smale diffeomorphism with prescribed periodic data were given by Shub and Sullivan.

(7.7) THEOREM [SS]. *If $f: M \longrightarrow M$ is Morse-Smale then there is a finitely generated free chain complex $C = \{C_k\}$ with chain map $\tau = \{\tau_k: C_k \longrightarrow C_k\}$ such that*

(1) *Each τ_k can be represented by a virtual permutation matrix.*

(2) *$H_*(C) \cong H_*(M)$ and under this isomorphism $\tau_*: H_*(C) \longrightarrow H_*(C)$ corresponds to $f_*: H_*(M) \longrightarrow H_*(M)$.*

(3) *The periodic data can be read off from the (signed) permutation blocks in the virtual permutation matrices for τ. In fact the ith such block in the matrix for τ_k is the companion matrix of $t^{p_{ik}} - \Delta_{ik}$.*

PROOF. Let $\{\gamma_i^k\}$, $1 \leqslant i \leqslant r(k)$, denote the basic sets of index k. According to (1.14) we can choose a Lyapunov function $g: M \longrightarrow R$ such that $g(\gamma_i^k) < g(\gamma_j^l)$ if $k < l$ (since the transversality condition then implies $W^u(\gamma_i^k) \cap W^s(\gamma_j^l) = \varnothing$). Renumbering if necessary we can also assume $g(\gamma_1^k) < g(\gamma_2^k) < \cdots < g(\gamma_{r(k)}^k)$ for all k.

We now define an open filtration of M, $\{X_i^k\}$, by setting $X_1^0 = W^s(\gamma_1^0)$, and letting $X_i^k = X_{i-1}^k \cup W^s(\gamma_i^k)$ and $X_1^{k+1} = X_{r(k)}^k \cup W^s(\gamma_1^{k+1})$. Thus $X_1^0 \subset X_2^0 \subset \cdots \subset X_{r(0)}^0 \subset X_1^1 \subset \cdots \subset X_{r(1)}^1 \subset X_1^2 \subset \cdots$ and if we let $M_k = X_{r(k)}^k$ then all basic sets of index k are contained in $M_k - M_{k-1}$.

We will show that $H_j(M_k, M_{k-1}) = 0$ unless $j = k$ and that $f_{*k}: H_k(M_k, M_{k-1}) \longrightarrow H_k(M_k, M_{k-1})$ can be represented by a virtual permutation matrix. It then follows by standard results that if $C_k = H_k(M_k, M_{k-1})$ and $\partial_k: C_k \longrightarrow C_{k-1}$ is the boundary map in the long exact sequence of the triple (M_k, M_{k-1}, M_{k-2}), $\{C_k, \partial_k\}$ is a chain complex whose homology is $H_*(M)$ (e.g. see (7.2) of [M1]). It is also easy to see that if we define $\tau_k = f_{*k}: C_k \longrightarrow C_k$ then τ_k is a chain map and τ_* is the map on $H_*(C) \cong H_*(M)$ induced by f. Thus to finish the proof we need only show $H_j(M_k, M_{k-1}) = 0$ if $j \neq k$ and that $f_{*k}: H_k(M_k, M_{k-1}) \longrightarrow H_k(M_k, M_{k-1})$ can be represented by a virtual permutation matrix. We will use the following lemma.

(7.8)   LEMMA.   *If $j \neq k$ then $H_j(X_i^k, X_{i-1}^k) = 0$ and $H_j(X_1^k, M_{k-1}) = 0$. Also $H_k(X_i^k, X_{i-1}^k) = Z^{p_{ik}}$ where p_{ik} is the period of γ_i^k and $f_{*k} : H_k(X_i^k, X_{i-1}^k) \to H_k(X_i^k, X_{i-1}^k)$ (or the map on $H_k(X_1^k, M_{k-1})$) can be represented by a matrix of the form*

$$
\begin{pmatrix}
0 & 1 & 0 & & 0 \\
 & 0 & 1 & 0 & \\
 & & & & \\
0 & & & & 1 \\
\Delta_{ik} & 0 & & & 0
\end{pmatrix}
.
$$

PROOF.   First we consider $Y_i^k = X_{i-1}^k \cup (\bigcup_j W_\epsilon^u(x_j))$, where $x_j \in \gamma_i^k$ and $W_\epsilon^s(x_j)$ is the disk of radius ϵ in $W^s(x_j)$ centered at x_j. We want to calculate $H_*(Y_i^k, X_{i-1}^k)$. Using the stable manifold theorem it is easy to see that there is a neighborhood U_j of x_j and coordinates $u_1, \ldots, u_n$ such that $W^s(x_j) \cap U_j$ is given by $u_1 = u_2 = \cdots = u_k = 0$ and $U_j \cap X_{i-1}^k = U_j - W^s(x_j)$. Hence one calculates easily that

$$
H_l(U_j \cap Y_i^k, U_j \cap X_{i-1}^k) = \begin{cases} Z & \text{if } l = k, \\ 0 & \text{if } l \neq k. \end{cases}
$$

Thus by excision

$$
H_l(Y_i^k, X_{i-1}^k) = \begin{cases} \bigoplus_j Z = Z^{p_{ik}} & \text{if } l = k, \\ \\ 0 & \text{if } l \neq k. \end{cases}
$$

Also we can choose a k-chain α_j lying in $W^u(x_j)$ such that α_j is a relative cycle and $\{[\alpha_j]\}$ represents a basis of $H_k(Y_i^k, X_{i-1}^k)$. If the orientations of the α_j's are chosen properly (cf. the proof of (7.6)) the matrix for $f_{*k} : H_k(Y_i^k, X_{i-1}^k) \to H_k(Y_i^k, X_{i-1}^k)$ will have the desired form. Finally we note $X_i^k = \bigcup_{n \geqslant 0} f^{-n}(Y_i^k)$ and for all $n > 0$ the inclusion $(f^{-n}(Y_i^k), X_{i-1}^k) \to (f^{-n-1}(Y_i^k), X_{i-1}^k)$ induces an isomorphism on homology. This gives the results claimed for $H_k(X_i^k, X_{i-1}^k)$. The results for $H_k(X_1^k, M_{k-1})$ are proved similarly.   Q.E.D.

We return now to the proof of (7.7). We will show by induction on j that $H_l(X_j^k, M_{k-1}) = 0$ if $l \neq k$ and that $f_{*k} : H_k(X_j^k, M_{k-1}) \to H_k(X_j^k, M_{k-1})$ is representable by a virtual permutation matrix. If $j = 1$ this follows from (7.8). We now assume the result for j and prove it for $j + 1$. We consider the exact sequence of the triple $(X_{j+1}^k, X_j^k, M_{k-1})$.

$$
H_{k+1}(X_{j+1}^k, X_j^k) \to H_k(X_j^k, M_{k-1}) \to H_k(X_{j+1}^k, M_k)
$$

$$
\to H_k(X_{j+1}^k, X_j^k) \to H_{k-1}(X_j^k, M_{k-1}).
$$

The two ends of this part of the sequence are 0 by (7.8) and the induction hypothesis. We wish to show f_{*k} on the middle term can be represented by a virtual permutation matrix. We know there are bases of $H_k(X_j^k, M_{k-1})$ and $H_k(X_{j+1}^k, X_j^k)$ such that f_{*k} with respect to each of them is a virtual permutation matrix. We choose a basis of $H_k(X_{j+1}, M_{k-1})$ which begins with the image of the basis of $H_k(X_j^k, M_{k-1})$ and extends it by a set which maps onto the basis of $H_k(X_{j+1}^k, X_j^k)$. With respect to this basis f_{*k} will be given by a virtual permutation matrix. The proof that $H_l(X_{j+1}^k, M_{k-1}) = 0$ if $l \neq k$ is similar.

Letting $j = r(k)$ we get $H_l(M_k, M_{k-1}) = 0$ if $l \neq k$ and that $f_{*k}: H_k(M_k, M_{k-1}) \to H_k(M_k, M_{k-1})$ is represented by a virtual permutation matrix. As remarked above this is all that is needed.   Q.E.D.

(7.9) COROLLARY [SS]. *If $f: M \to M$ is Morse-Smale then all eigenvalues of f_*: $H_*(M; R) \to H_*(M; R)$ are roots of unity.*

PROOF.   All eigenvalues of a virtual permutation matrix are roots of unity. Any eigenvalue of $f_{*k}: H_k(M) \to H_k(M)$ must also be an eigenvalue of $\tau_k: C_k \to C_k$, so the result follows from (7.7). This result also follows easily from (7.3) by induction on q.   Q.E.D.

There is a converse to (7.7) when M satisfies certain additional hypotheses.

(7.10) THEOREM [SS]. *Suppose M is a simply connected compact manifold of dimension greater than five with torsion free homology and $f: M \to M$ is a diffeomorphism. Then a sufficient condition that there exist a Morse-Smale diffeomorphism with periodic data $\{(p_{ik}, \Delta_{ik})\}$ is that there exist a finitely generated free chain complex C with $C_1 = C_{n-1} = 0$ and an endomorphism $\tau: C \to C$ such that $H_*(C) \cong H_*(M)$ and under this isomorphism $\tau_*: H_*(C) \to H_*(C)$ corresponds to $f_*: H_*(M) \to H_*(M)$, and such that, for each $1 \leq k \leq n$, $\tau_k: C_k \to C_k$ is representable by a virtual permutation matrix*

$$\begin{pmatrix} P_1 & * & & & * \\ 0 & P_2 & * & & \\ & & 0 & & \\ & & & & * \\ 0 & & 0 & & P_r \end{pmatrix}$$

where P_i is the companion matrix of $t^{p_{ik}} - \Delta_{ik}$, i.e. has the form

$$\begin{pmatrix} 0 & 1 & 0 & & 0 \\ & 0 & 1 & 0 & \\ & & & & \\ 0 & & & & 1 \\ \Delta_{ik} & 0 & & & 0 \end{pmatrix}.$$

PROOF. This is an immediate consequence of (4.7) and (7.6).

We can use this result to show that the necessary conditions for the existence of a Morse-Smale diffeomorphism with prescribed periodic data given by the Morse inequalities (7.3) are often sufficient. This is an improvement over (7.10) since it is much easier to check conditions on $f_*: H_*(M) \rightarrow H_*(M)$ than conditions on some unknown chain level representation of f_*.

(7.11) THEOREM [FN]. *Suppose $f: M \rightarrow M$ is a diffeomorphism of a simply connected manifold of dimension $n > 5$, with torsion free homology. Given periodic data $\{(p_{ik}, \Delta_{ik})\}$ without orbits of index 1 or $n - 1$ and which satisfy the inequalities (7.3) for f, a sufficient condition that there exist a Morse-Smale diffeomorphism of M isotopic to f realizing these data is that for each k there exist monic integer polynomials $\{g_{ik}(t)\}$ such that g_{ik} is a factor of $(t^{p_{ik}} - \Delta_{ik})$ and the map $f_{*k}: H_k(M) \rightarrow H_k(M)$ is representable by a matrix of the form*

$$
\begin{pmatrix}
C(g_{1k}) & & * & & & * \\
0 & C(g_{2k}) & & * & & \\
& & 0 & & & \\
& & & & & * \\
0 & & & 0 & & C(g_{lk})
\end{pmatrix}.
$$

where $C(g_{ik})$ denotes the companion matrix of g_{ik}. In particular it is sufficient if f is homotopic to the identity.

Prior to giving the proof of (7.11) we need two preliminary results.

(7.12) LEMMA [FN]. *Suppose for each q, $f: M \rightarrow M$ and the periodic data $\{(p_{ik}, \Delta_{ik})\}$ satisfy the conclusion of (7.3) for some choice of P_q. Then for each q there is a polynomial $h_q(t)$ such that*

$$
\prod_i (t^{p_{iq}} - \Delta_{iq}) = h_q h_{q+1} \det(tI - f_{*q}).
$$

PROOF. By the conclusion of (7.3) we have that

$$
\prod_{k \leqslant q} (1 - \Delta_{ik} t^{p_{ik}})^{(-1)^k} = P_q^{(-1)^q} \cdot \prod_{k \leqslant q} \det(I - f_{*k} t)^{(-1)^k}.
$$

Dividing this equality by the same equality with q replaced by $(q - 1)$ we obtain

$$
\prod_i (1 - \Delta_{iq} t^{p_{iq}}) = P_q P_{q-1} \det(I - f_{*q} t).
$$

Now on both sides of this equation we replace each factor $g(t)$ by $g(1/t) \cdot t^{\deg g}$. Thus for example $\det(I - f_{*q} t)$ is replaced by $t^m \det(I - f_{*q} t^{-1}) = \det(tI - f_{*q})$ and we obtain

$$\prod_i (t^{p_{iq}} - \Delta_{iq}) = h_q h_{q+1} \det(tI - f_{*q})$$

where $h_{q+1} = P_q(t^{-1}) t^{\deg P_q}$.

We also need a fact from **[FN]** concerning similarity classes over the integers of companion matrices. We will denote by $C(p)$ the companion matrix of the monic polynomial $p(x)$ and the $n \times m$ matrix with 1 in the nth row and 1st column and 0 elsewhere will be denoted by $E(n, m)$.

(7.13) PROPOSITION **[FN]**. *Let $p(x)$ be a monic polynomial with integer coefficients. If $p(x) = f_1(x) f_2(x) \cdots f_k(x)$, then $C(p)$ is similar over the integers to the block triangular matrix.*

$$\begin{pmatrix} C(f_1) & E(d_1 d_2) & 0 & \cdots & 0 \\ 0 & C(f_2) & E(d_2 d_3) & \cdots & 0 \\ \vdots & & & & \\ 0 & 0 & 0 \cdots & C(f_{k-1}) & E(d_{k-1} d_k) \\ 0 & 0 & 0 \cdots & 0 & C(f_k) \end{pmatrix},$$

where $d_i = \deg f_i(x)$.

This result may be of some independent algebraic interest. However, its proof is rather lengthy and unenlightening, so we therefore allow the interested reader to consult (2.3) of **[FN]**.

PROOF OF (7.11). By the theorem of Shub and Sullivan, (7.10) above, it suffices to construct a chain complex together with a chain map $\tau: C \to C$ such that $H_*(C) = H_*(M)$, $\tau_* = f_*$ and, for each k, $\tau_k: C_k \to C_k$ has the form

(1)
$$\begin{pmatrix} P_1 & * & & * \\ 0 & P_2 & * & \\ & 0 & & \\ & & & * \\ 0 & & 0 & P_r \end{pmatrix}$$

where $P_i = C(t^{p_{ik}} - \Delta_{ik})$ is the $p_{ik} \times p_{ik}$ matrix determined by the periodic data $\{(p_{ik}, \Delta_{ik})\}_{i=1}^{r(k)}$ for the orbits of index k.

By (7.12) there exist polynomials h_k, $0 \leqslant k \leqslant n$, such that

$$\prod_{i=1}^{r} (t^{p_{ik}} - \Delta_{ik}) = h_k h_{k+1} \det(tI - f_{*k}) = h_k h_{k+1} \prod_{i=1}^{r} g_{ik}.$$

Thus we can choose monic polynomials a_{ik} and b_{ik} such that $t^{p_{ik}} - \Delta_{ik} = a_{ik} g_{ik} b_{ik}$ for all i (some of the a_{ik}, b_{ik} and g_{ik} may be identically 1) and such that

$$\prod_{i=1}^{r} a_{ik} = h_k, \qquad \prod_{i=1}^{r} b_{ik} = h_{k+1}.$$

We now form a square integer matrix τ_k' of size $\deg h_k + \deg h_{k+1} + \deg(\det(tI - f_{*k}))$ and let C_k be the free abelian group of this rank. Let τ_k' be given by

$$\begin{pmatrix} B_k & F_k & 0 \\ 0 & G_k & F_k' \\ 0 & 0 & A_k \end{pmatrix}$$

where the entries here are matrices defined as follows: The matrix B_k is given by

$$\begin{pmatrix} C(b_{1k}) & S_1 & 0 & & 0 \\ 0 & C(b_{2k}) & S_2 & 0 & \\ & 0 & & & 0 \\ & & & & S_{r-1} \\ 0 & & 0 & & C(b_{rk}) \end{pmatrix}$$

where S_i is the matrix of size $\deg(g_{ik})$ by $\deg(g_{(i+1)k})$ of the form

(2)
$$\begin{pmatrix} 0 & \cdots 0 \\ 0 & \\ 1 & 0 \cdots 0 \end{pmatrix}.$$

Likewise A_k is defined to be

$$
\begin{pmatrix}
C(a_{1k}) & S_1' & 0 & & & \\
0 & C(a_{2k}) & S_2' & 0 & & \\
& & & & 0 & \\
& & & & S_{r-1}' & \\
0 & & & 0 & & C(a_{rk})
\end{pmatrix}
$$

where the form of S_i' is the same as that of S_i. The matrix G_k is the matrix representing $f_{*k} \colon H_k(M) \longrightarrow H_k(M)$ whose existence is given by the hypothesis. Finally the matrices F_k and F_k' both have the form

$$
\begin{pmatrix}
0 & T_1 & & 0 \\
& 0 & T_2 & \\
& & 0 & \\
& & & T_{r-1} \\
0 & & & 0
\end{pmatrix}
$$

where T_i has the same form (2) as S_i and S_i' but its size is $\deg(b_{ik})$ by $\deg(g_{ik})$ for F_k and $\deg(g_{ik})$ by $\deg(a_{ik})$ for F_k'.

We proceed now to define the boundary map $\partial_k \colon C_k \longrightarrow C_{k-1}$ for the chain complex C. Let Z_k be the summand of C_k consisting of elements whose only nonzero coordinates are among the first $\dim B_k + \dim G_k$ coordinates, that is, Z_k is the summand invariant under τ_k' and with the action of τ_k' being given by the matrix

$$
\begin{pmatrix}
B_k & F_k \\
0 & G_k
\end{pmatrix}
$$

We similarly define D_k to be the τ_k' invariant summand of C_k consisting of elements whose only nonzero components are among the first $\dim B_k$ so that the action of τ_k' on D_k is given by the matrix B_k. Note that the induced map $\overline{\tau}_k \colon C_k/Z_k \longrightarrow C_k/Z_k$ is represented by the matrix A_k.

We can now define the boundary map $\partial_k \colon C_k \longrightarrow C_{k-1}$ by specifying that Z_k is its kernel and choosing an isomorphism for the induced map $\overline{\partial}_k \colon C_k/Z_k \longrightarrow D_{k-1} \subset C_{k-1}$ so that the diagram

$$\begin{array}{ccc}
C_k/Z_k & \xrightarrow{\ A_k\ } & C_k/Z_k \\
\downarrow{\scriptstyle \bar{\partial}_k} & & \downarrow{\scriptstyle \bar{\partial}_k} \\
D_{k-1} & \xrightarrow{\ B_{k-1}\ } & D_{k-1}
\end{array}$$

commutes. This is possible because, as a consequence of (7.13), the matrices A_k and B_{k-1} are similar over the integers since $\Pi_i a_{ik} = h_k = \Pi_j b_{j(k-1)}$.

It is clear that we have defined a chain complex $C = \{C_k\}$ and a chain map τ'_k: $C_k \longrightarrow C_k$. Also the induced map $\tau'_{*k}: H_k(C) = Z_k/D_k \longrightarrow Z_k/D_k$ is represented by the matrix G_k. Thus to complete the proof we need only show that τ'_k is similar over Z to a matrix τ_k of the form (1).

If we permute basis elements of C_k in such a way as to put the diagonal blocks in the order $C(b_{1k})$, $C(g_{1k})$, $C(a_{1k})$, $C(b_{2k})$, $C(g_{2k})$, $C(a_{2k})$, $C(b_{3k})$, etc., we obtain a matrix τ''_k of the form

$$\begin{pmatrix}
Q_1 & * & & * \\
0 & Q_2 & * & \\
& & & * \\
0 & & 0 & Q_r
\end{pmatrix}$$

where Q_i has the form

$$\begin{pmatrix}
C(b_{ik}) & S & 0 \\
0 & C(g_{ik}) & S' \\
0 & 0 & C(a_{ik})
\end{pmatrix}$$

and S and S' have all entries zero except a 1 in the last entry of their first columns.

Since $b_{ik} g_{ik} a_{ik} = t^{p_{ik}} - \Delta_{ik}$ another application of (7.13) shows that Q_i is similar over the integers to $P_i = C(t^{p_{ik}} - \Delta_{ik})$. Hence the matrix τ''_k is similar over the integers to a matrix τ_k of the form (1).   Q.E.D.

We turn now to the more basic question of whether a given homotopy class admits *any* Morse-Smale diffeomorphism. By (7.9), a necessary condition is that all the eigenvalues of $f_*: H_*(M) \longrightarrow H_*(M)$ be roots of unity. Theorem (7.10) gives good sufficient conditions in terms of chain level representations of f_* and (7.11) can be used to show f is isotopic to a Morse-Smale diffeomorphism provided f_* has the special form of its hypothesis.

A systematic algebraic answer to this question of existence in terms of $f_*: H_*(M) \longrightarrow H_*(M)$ for simply connected manifolds of dimension > 5 was given in [F-Sh]. In addition to the condition on the eigenvalues there is an obstruction involving the algebraic K theory of endomorphisms whose eigenvalues are roots of unity or zero. Before stating the theorem we give a brief account of this.

An endomorphism $e: F \rightarrow F$ of a finitely generated abelian group F is called *quasi-idempotent* provided the induced endomorphism on $F/$torsion has only eigenvalues which are roots of unity or zero.

(7.14) DEFINITION. Let QI denote the category of quasi-idempotent endomorphisms of finitely generated abelian groups.

Thus, an object in QI is a pair (F, e) where F is a finitely generated abelian group and e is a quasi-idempotent endomorphism of F. A morphism $h: (F_1, e_1) \rightarrow (F_2, e_2)$ in QI is a homomorphism $h: F_1 \rightarrow F_2$ such that $h \circ e_1 = e_2 \circ h$. An isomorphism in QI is defined similarly. A short exact sequence in QI is a sequence of morphisms $0 \rightarrow (F_1, e_1) \rightarrow (F_2, e_2) \rightarrow (F_3, e_3) \rightarrow 0$ which is exact on the level of abelian groups.

The additional obstruction to the existence of a Morse-Smale diffeomorphism lies in the torsion subgroup of the group $K_0(QI)$, which we now define (see [Bass1] for more details).

(7.15) DEFINITION. The Grothendieck group $K_0(QI)$ is the abelian group with generators $\{[(F, e)]\}$, the isomorphism classes of elements of QI, and with relations

$$[(F_2, e_2)] = [(F_1, e_1)] + [(F_3, e_3)]$$

whenever there is an exact sequence $0 \rightarrow (F_1, e_1) \rightarrow (F_2, e_2) \rightarrow (F_3, e_3) \rightarrow 0$.

Given an element (F, e) in QI we denote its class in $K_0(QI)$ by $[(F, e)]$.

(7.16) PROPOSITION [F-Sh]. *The group $K_0(QI)$ is isomorphic to $P \oplus G$ where G is the torsion subgroup of $K_0(QI)$ and P is the multiplicative group of rational functions whose numerators and denominators are monic polynomials with roots either zero or roots of unity.*

PROOF. We first prove that every element of $K_0(QI)$ can be written in the form $[(F_1, e_1)] - [(F_2, e_2)]$ where F_1 and F_2 are free abelian groups. It suffices to show this for elements of the form $[(T, e)]$ where T is a finite group. Let ZT be the free abelian group with T as a set of generators and $\rho: ZT \rightarrow T$ the homomorphism given by $\rho(ng) = g + \cdots + g$ (n times). Let $\hat{e}: ZT \rightarrow ZT$ be the endomorphism induced by the action of e on generators so $\rho: (ZT, \hat{e}) \rightarrow (T, e)$ is a morphism in QI. If $K = \ker \rho$ then $0 \rightarrow (K, \hat{e}|K) \rightarrow (ZT, \hat{e}) \rightarrow (T, e) \rightarrow 0$ is exact and K is free so

$$[(T, e)] = [(K, \hat{e}|K)] - [(ZT, \hat{e})].$$

We now define a homomorphism $p: K_0(QI) \rightarrow P$ by

$$p([(F_1, e_1)] - [(F_2, e_2)]) = f_1(t)/f_2(t)$$

where F_i is free and $f_i(t)$ is the characteristic polynomial of e_i. It is easy to check that this is well defined. Let the map $i: P \rightarrow K_0(QI)$ be given by $i(f_1(t)/f_2(t)) = [(F_1, C(f_1))] - [(F_2, C(f_2))]$ where $C(f_i)$ is the companion matrix of f_i and F_i is a free abelian group of rank $\deg f_i$. Then i is a homomorphism and a right inverse to p, i.e., $p \circ i = \mathrm{id}: P \rightarrow P$.

Thus $K_0(QI) \cong P \oplus \ker(p)$ so it will suffice to show that $\ker(p)$ contains only elements of finite order and hence is isomorphic to the torsion subgroup G of $K_0(QI)$. An element $[(F, e)] - [(F', e')]$ with F and F' free is in $\ker(p)$ if and only if the characteristic polynomials of e and e' are the same.

It is a well-known fact (cf. [Ne, p. 50]) that we can choose a basis for F such that the matrix A corresponding to e has the form

$$
\begin{pmatrix}
A_1 & * & & * \\
 & \cdot & & \\
0 & & \cdot & \\
 & & & * \\
 & & \cdot & \\
0 & & 0 & A_n
\end{pmatrix}
$$

where each A_i has an irreducible characteristic polynomial. It follows that in $K_0(QI)$, $[e] = \Sigma_i [e_i]$ where e_i is represented by the matrix A_i (here and henceforth we suppress the group F and simply write $[e]$ in place of the more cumbersome $[(F, e)]$). Also if $([e] - [e']) \in \ker(p)$, $[e'] = \Sigma_i [e_i']$ where each e_i' has irreducible characteristic polynomial g_i equal to the characteristic polynomial of e_i. To prove that $([e] - [e']) \in \ker(p)$ has finite order it is sufficient to show that $n[e_i] = n[c_i] = n[e_i']$ for some n, where c_i is an endomorphism of a free abelian group given by the companion matrix $C(g_i)$ of g_i.

If $[e_i]$ and $[e_i']$ are the zero endomorphism then this is trivial so we may assume that g_i is a cyclotomic polynomial Φ_k.

Let ω be a primitive kth root of unity and let $\Lambda = Z[\omega]$. Isomorphism classes of ideals in the Dedekind domain Λ form a group under multiplication called the ideal class group. The ideal class of Λ itself, i.e., the class of principal ideals, is the group identity element. This group is finite (see [BS]). Isomorphism classes of elements of QI with characteristic polynomial Φ_k are in one-to-one correspondence with the elements of the ideal class group of Λ (see for example [Ne, p. 53]). The correspondence is given as follows: If I is an ideal in Λ then it is a free Z module and multiplication by ω is a Z linear endomorphism and hence determines a similarity class of matrices with characteristic polynomial Φ_k. It is easy to see that the principal ideal class, i.e., the class of Λ, corresponds to the companion matrix $C(g_i)$.

Given ideals $I_1, \ldots, I_m$ in Λ, $\bigoplus_j I_j \simeq \Lambda^{m-1} \oplus I_1 I_2 \cdots I_m$ as Λ modules (see (1.6) of [M2]). Given an ideal I, we apply this with $I_j = I$ for all j, and m the order of the element of the ideal class group determined by I, to obtain $I^m \cong \Lambda^m$ as Λ modules.

If I is an ideal in the ideal class corresponding to e_i (i.e., if the endomorphism of I given by multiplying by ω is similar to e_i) and if $h: I^m \longrightarrow I^m$ is the endomorphism obtained by multiplying by ω, then clearly $[h] = m[e_i]$. But since $I^m \cong \Lambda^m$ as Λ modules, we also have $[h] = m[C(g_i)]$, so $m[e_i] = m[C(g_i)]$. A similar argument shows $m'[e_i'] = m'[C(g_i)]$ so if $n = mm'$, we have $n[e_i] = n[C(g_i)] = n[e_i']$. Thus $[e_i] - [e_i']$ has finite order.   Q.E.D.

(7.17) DEFINITION. Let $\phi\colon K_0(QI) \longrightarrow G$ be the composite

$$K_0(QI) \cong P \oplus G \longrightarrow G$$

where the first map is the isomorphism given by (7.16) and the second map is projection onto G.

(7.18) LEMMA. *If $(F, e) \in QI$, with F free, and e is representable by a companion matrix, then $\phi([e]) = 0 \in G$.*

PROOF. It is clear from the proof of (7.16) that the isomorphism $K_0(QI) \cong P \oplus G$ is given by

$$K_0(QI) \xrightarrow{(p,\phi)} P \oplus G$$

where, as above, p assigns to $[e]$ its characteristic polynomial. Since we showed in the proof of (7.16) that $p \circ i = $ id: $P \longrightarrow P$ it follows that if $(F, e) \in QI$ and e is representable by a companion matrix then $\phi([e]) = 0 \in G$.   Q.E.D.

(7.19) THEOREM [F-Sh]. *Suppose $f\colon M \longrightarrow M$ is a diffeomorphism and $f_*\colon H_*(M; Z) \longrightarrow H_*(M; Z)$ has only eigenvalues which are roots of unity. Then if f is homotopic to a Morse-Smale diffeomorphism, $\chi(f_*) = \Sigma(-1)^k \phi([f_{*k}])$ is zero in the torsion subgroup G of $K_0(QI)$. If M is simply connected and of dimension greater than five, then $\chi(f_*) = 0$ implies that f is isotopic to a Morse-Smale diffeomorphism.*

We give the proof of only the first part of this theorem. The interested reader is referred to [F-Sh] for the second half. Thus we want to show that if f is Morse-Smale $\chi(f_*) = 0$. We can define a map α from the objects of QI to G by $\alpha(e) = \phi([e])$. The fact that α depends only on the equivalence class of e, $[e]$, in $K_0(QI)$ and the fact that ϕ is a homomorphism imply that α is additive on short exact sequences (i.e., it is an Euler-Poincaré function). Thus we can apply (5.9) and obtain

$$\chi(f_*) = \sum(-1)^k \alpha(\tau_k) = \sum(-1)^k \phi([\tau_k])$$

if, for each k, $\tau_k\colon C_k \longrightarrow C_k$ is a chain endomorphism which is in QI and which represents $f_*\colon H_*(M) \longrightarrow H_*(M)$ on the chain level.

However, by (7.7) if f is Morse-Smale there is such a chain map $\tau_k\colon C_k \longrightarrow C_k$ with each τ_k representable by a virtual permutation matrix

$$\begin{pmatrix} P_{ik} & * & * \\ & \ddots & * \\ 0 & & P_{mk} \end{pmatrix}$$

where m depends on k. Thus $\phi([\tau_k]) = \Sigma_j \phi([P_{jk}]) = 0$ by (7.18). It follows that $\chi(f_*) = \Sigma_k(-1)^k \phi([\tau_k]) = 0$.   Q.E.D.

The torsion subgroup G of $K_0(I)$ has been computed through the work of H. Bass, D. Grayson and H. Lenstra (see [Bass2]). It is a very large group containing for example infinitely many copies of $Z/p^n Z$ for every $n \geqslant 1$ and every odd prime p.

Techniques from Chapter 4 and Appendix B can be used to show that each element G can be realized as $\chi(f_*)$ for some diffeomorphism $f: M \longrightarrow M$ of a manifold with boundary. If M is a closed manifold of dimension n, then because of Poincaré duality, $\chi(f_*)$ will lie in the (perhaps) smaller group $\bar{G}$ generated by elements of G of the form $\phi([e] + (-1)^n [\bar{e}])$, where e is an automorphism in QI and $\bar{e}$ is its inverse transpose. (I am grateful to W. Browder for this remark.) More algebraic work might be valuable here.

Given a particular endomorphism e, it does not seem to be clear how to find $\phi([e])$ or even check whether or not it is nonzero. Fortunately in practice the need may never arise as the following theorem shows.

(7.20) THEOREM. *Suppose $f: M \longrightarrow M$ is a diffeomorphism, and all eigenvalues of $f_*: H_*(M; Z) \longrightarrow H_*(M; Z)$ are roots of unity, i.e., if λ_i is an eigenvalue then $\lambda_i^{n_i} = 1$ for some n_i. If $n_i < 23$ for all i, and the order of the torsion subgroup of $H_k(M; Z)$ is < 23 for all k, then $\chi(f_*) = 0$.*

PROOF. We first consider the case that $H_*(M)$ is torsion free. We know [Ne, p. 50] that with respect to some basis a matrix for $f_{*k}: H_k(M) \longrightarrow H_k(M)$ will have the form

$$\begin{pmatrix} A_1 & * & & & * \\ & \cdot & & & \\ 0 & & \cdot & & \\ & & & \cdot & * \\ 0 & & 0 & & A_n \end{pmatrix}$$

where each A_j has an irreducible characteristic polynomial which must be a cyclotomic polynomial. We again use the correspondence between similarity classes of matrices whose characteristic polynomial is the cyclotomic polynomial Φ_n with ideal classes of $Z[\omega]$, where ω is a root of Φ_n (see the proof of (7.16) and [Ne, p. 53]). Since in this case $n < 23$ and the ideal class group of $Z[\omega]$ is trivial when $n < 23$ (see [BS]) we conclude each A_j is similar to a companion matrix. It follows then from (7.18) that

$$\phi([f_{*k}]) = \sum_j \phi([A_j]) = 0,$$

so $\chi(f_*) = 0$.

Now if T is the torsion subgroup of $H_*(M; Z)$ and $e: T \longrightarrow T$ is the endomorphism induced by f_{*k}, then as we showed in the beginning of the proof of (7.16), $[(T, e)] = [(K, \hat{e}|K)] - [(ZT, \hat{e})]$. The endomorphism $\hat{e}$ can be represented by a permutation matrix of size $< 23 \times 23$. Hence the roots of its characteristic polynomial (which is the same as that of $\hat{e}|K$) are roots of unity λ_i with $\lambda_i^{n_i} = 1$ for some $n_i < 23$. It follows that the same argument given above implies $\phi([\hat{e}]) = 0$ and $\phi([\hat{e}|K]) = 0$ so $\phi([e]) = 0$. Thus we again have $\phi([f_{*k}]) = 0$ so $\chi(f_*) = 0$.   Q.E.D.

(7.21) *Exercises.* (1) (*D. Fried*). Show that if $f\colon M \longrightarrow M$ is a diffeomorphism of a simply connected closed manifold and all eigenvalues of $f_*\colon H_*(M; R) \longrightarrow H_*(M; R)$ are roots of unity, then $f \times \mathrm{id}\colon M \times S^{2n+1} \longrightarrow M \times S^{2n+1}$, $n \geqslant 1$ is isotopic to a Morse-Smale diffeomorphism.

(2) [SS]. Show that if $f\colon M \longrightarrow M$ is a diffeomorphism of a simply connected manifold of dimension > 5, and all the eigenvalues of $f_*\colon H_*(M) \longrightarrow H_*(M)$ are roots of unity, then f^n is isotopic to a Morse-Smale diffeomorphism for some $n \geqslant 1$.

Chapter 8. Morse-Smale flows

We turn now to the study of flows with a hyperbolic structure on their chain recurrent set. We have already encountered some of the simplest examples of such flows;
namely, flows generated by integrating the gradient of a Morse function—the gradient-like
flows of Chapter 2. The chain recurrent set of these flows is the finite set of critical
points of the associated Morse function. It is natural to enlarge this class somewhat to
include flows whose chain recurrent set R consists of a finite set of orbits, so that R will
include a finite set of periodic orbits as well as rest points.

(8.1) DEFINITION. A smooth flow f_t on M is called *Morse-Smale* provided

(a) The chain recurrent set R of f_t consists of a finite number of hyperbolic closed
orbits and hyperbolic rest points.

(b) The unstable manifold of any closed orbit or rest point has transversal intersection
with the stable manifold of any closed orbit or rest point.

Clearly a Morse-Smale flow has a hyperbolic chain recurrent set and satisfies the transversality condition (1.9) and hence is structurally stable (1.10).

As in our study of diffeomorphisms we want first to provide the means of systemati-
cally producing examples and then to obtain homological restrictions on the kinds of dynamics
which can occur. For nonsingular Morse-Smale flows the examples we can construct and the
restrictions homology imposes do meet and in Theorem (8.9) we give a necessary and sufficient conditions theorem analagous to the Morse inequality theorem for gradient-like flows
(2.3).

(8.2) DEFINITION. A hyperbolic closed orbit γ of a flow will be called *twisted* if its
unstable bundle $E^u(\gamma)$ is unorientable and *untwisted* if this bundle is orientable.

Thus if γ is an untwisted closed orbit of index k, $W^u(\gamma)$ is diffeomorphic to $S^1 \times R^k$
and if γ is twisted $W^u(\gamma)$ is diffeomorphic to (open Moebius strip) $\times R^{k-1}$.

The basic building block for constructing flows with closed orbits is the following result of Asimov.

(8.3) THEOREM [As1]. *Suppose that* $\psi\colon (M, V, W) \to ([a, b], a, b)$ *(where*
$\partial M = V \cup W$) *is a Morse function with critical points p and q of index k and $k + 1$*
respectively. If $W^s(p) \cap W^u(q) = \emptyset$ *then there is a Morse-Smale flow on M, agreeing*
with $-\nabla\psi$ *on ∂M, whose chain-recurrent set R consists of one untwisted closed orbit of*
index k.

The importance of this result lies in the fact that it reduces the problem of constructing examples of Morse-Smale flows to one of constructing Morse functions with certain kinds of critical points and this is a fairly well-understood problem, at least for simply connected manifolds of dimension > 5 (see Chapter 2, especially (2.17)).

SKETCH OF PROOF. We note that if $\psi(p) < c < \psi(q)$ and $L = \psi^{-1}(c)$ then $W^u(q) \cap L$ is a k dimensional sphere in L (see the remarks after (2.7)). Likewise $W^s(p) \cap L$ is an $(n - k - 1)$ dimensional sphere. We denote these by $S^u(q)$ and $S^s(p)$ respectively. The first step is to alter ψ so that for the new flow $S^s(p) \cap S^u(q)$ consists of two points with opposite intersection numbers (initially $S^u(p)$ and $S^u(q)$ are disjoint).

Then using techniques and ideas from (5.5) of [M1] it is possible to alter the flow on a neighborhood of the two trajectories running from q to p, i.e., the trajectories through $S^s(p) \cap S^u(q)$, in such a way that these two trajectories for the new flow form a smooth circle through p and q. In fact it is possible to arrange that there are local coordinates $(\vec{x}, \vec{y}, \theta)$, $\vec{x} \in R^k$, $\vec{y} \in R^{n-k-1}$, $\theta \in S^1$ on a neighborhood U of these trajectories and in these coordinates the vector field tangent to the new flow is given by

$$X = \sin \theta \frac{\partial}{\partial \theta} + \sum_{i=1}^{k} x_i \frac{\partial}{\partial x_i} - \sum_{j=1}^{n-k-1} y_j \frac{\partial}{\partial y_j}.$$

If U is chosen small enough and the crrect shape then no trajectory leaving U will ever re-enter it.

We now choose a smooth monotonic function $\rho(t)$ and $\epsilon > 0$ so that $|(\vec{x}, \vec{y})| < \epsilon$ implies $(\vec{x}, \vec{y}, \theta) \in U$ and with $\rho(t) = 0$ if $t > \epsilon$, $\rho(t) = 1$ if $t < \epsilon/2$. Then define

$$X' = [\rho(|\vec{x}, \vec{y}|)(1 - \sin \theta)] \frac{\partial}{\partial \theta} + X$$

so that $X = X'$ outside U and $X' = \partial/\partial\theta + \Sigma x_i \partial/\partial x_i - \Sigma y_j \partial/\partial y_j$ on a neighborhood of the circle $\vec{x} = 0$, $\vec{y} = 0$. Clearly the flow generated by X' has the desired closed orbit. Also if $g(\vec{x}, \vec{y}) = -\Sigma x_i^2 + \Sigma y_j^2$ then $X'(g) \leq 0$ with equality only on the closed orbit. It follows from this and the fact that nothing exiting U ever re-enters that R consists of only this one orbit. The transversality condition can be obtained by a further perturbation according to the Kupka-Smale theorem [S1]. Q.E.D.

(8.4) REMARK. We could just as well have started with a ψ such that $W^u(q) \cap W^s(p)$ consisted of two trajectories. If the two points of $S^u(q) \cap S^s(p)$ have opposite intersection numbers then we are in the situation above. If not a similar construction can be performed yielding a *twisted* closed orbit.

The reverse construction which replaces a closed orbit of index k with a rest point of index k and one of index $(k + 1)$ works also.

(8.5) PROPOSITION [F5]. *Suppose f_t is a Morse-Smale flow with a closed orbit γ of index k. Then given a neighborhood U of γ there exists a new Morse-Smale flow g_t whose vector field agrees with that of f_t outside U and which has rest points p, q of index k and $k + 1$ in U but no other chain recurrent points in U. For g_t, $S^u(q) \cap S^s(p)$ will consist of*

*two points with the same intersection number if γ was twisted and opposite intersection
numbers if γ was untwisted.*

SKETCH OF PROOF. We first alter f_t so that on a small neighborhood U its vector
field has the form

$$X = \frac{\partial}{\partial \theta} + \sum_{i=1}^{k} x_i \frac{\partial}{\partial x_i} - \sum_{j=1}^{n-k-1} y_j \frac{\partial}{\partial y_j}$$

in local coordinates $(\vec{x}, \vec{y}, \theta) \in R^k \times R^{k-1} \times S^1$. If we then shrink U keeping it the correct shape we can guarantee that no point leaving U will later re-enter it. We now reverse
the construction of (8.3) to obtain a flow whose vector field on a still smaller neighborhood
of γ is given by

$$X' = ((1 - \rho) + \rho \sin \theta)\frac{\partial}{\partial \theta} + \sum x_i \frac{\partial}{\partial x_i} - \sum y_j \frac{\partial}{\partial y_j}$$

where $\rho = \rho(|(\vec{x}, \vec{y})|)$ is as in (8.3). If $\phi(\vec{x}, \vec{y}, \theta) = -|\vec{x}|^2 + |\vec{y}|^2 - \cos \theta$ then $X'(\phi) \leqslant 0$
with equality only at the two rest points $p = (\vec{0}, \vec{0}, 0)$ and $q = (\vec{0}, \vec{0}, \pi)$. This is sufficient
to guarantee that the only chain recurrent points in U are p and q. The transversality condition comes from using the Kupka-Smale theorem [S1] and perturbing f_t again.   Q.E.D.

The results (8.3) and (8.5) show that a closed orbit is to a certain extent interchange-
able with a pair of rest points p and q with the property of being twisted or untwisted corresponding to the way $W^u(q)$ and $W^s(p)$ intersect.

An easy consequence of this discussion is the Morse inequalities of Smale for Morse-
Smale flows.

(8.6)   THEOREM [S6]. *Suppose f_t is a Morse-Smale flow on a compact manifold M
with c_k rest points of index k and A_k closed orbits of index k. Then*

$$c_k - c_{k-1} + \cdots \pm c_0 + A_k \geqslant \beta_k - \beta_{k-1} + \cdots \pm \beta_0$$

where $\beta_k = \dim H_k(M; F)$ and F is a field.

PROOF. Form a new flow by using (8.5) to replace each closed orbit with two critical
points. The new flow can be altered on a neighborhood of its critical points so that it is
gradient-like in the sense of Chapter 2 without changing the numbers or indices of critical
points. This new flow has c_k' critical points of index k where $c_k' = c_k + A_k + A_{k-1}$. Applying (2.12) to this flow gives the desired inequalities.   Q.E.D.

REMARK. If the field F does not have characteristic 2 the above result can be strength-
ened. In fact the same inequalities with A_k replaced by A_k^u = (number of untwisted closed
orbits of index k) are valid in this case. The proof of this stronger version of the result is
an exercise in the use of long exact sequences and induction. We will give a proof of a more
general result in (9.12).

We turn now to the study of nonsingular Morse-Smale flows, i.e., Morse-Smale flows
which have no rest points. The question of existence of such flows was largely resolved by
Asimov.

(8.7) THEOREM [As2]. *If M is a compact manifold of dimension $\geqslant 4$ with zero Euler characteristic, then any homotopy class of nonsingular vector fields on M contains a vector field generating a nonsingular Morse-Smale flow.*

The proof, which we omit, relies on repeated application of (8.3). Consequently the flow constructed has no twisted closed orbits. The situation for three dimensional manifolds is quite different as the following result of Morgan shows.

(8.8) THEOREM [Mo]. *There exist compact three dimensional manifolds which do not admit any nonsingular Morse-Smale flow.*

In fact Morgan gives a characterization of those three-manifolds which do admit a non-singular Morse-Smale flow.

After the question of existence the next natural question to pose is what kinds of periodic behavior can occur on a manifold with a given homological configuration. Can we give necessary and sufficient conditions analagous to Theorem (2.18) for Morse functions? The following result comes close to realizing this result for nonsingular Morse-Smale flows but it does not address the possibility of twisted closed orbits.

(8.9) THEOREM [F5]. *Suppose that M^n admits a nonsingular Morse-Smale flow with A_k untwisted closed orbits and $\beta_k = \dim H_k(M; R)$. Then*

(a) $A_k \geqslant \beta_k - \beta_{k-1} + \cdots \pm \beta_0$ *for all k,*

(b) $A_1 \geqslant A_0 - 1$ *and* $A_{n-2} \geqslant A_{n-1} - 1$, *and*

(c) *If* $A_{k-1} = A_{k+1} = 0$ *and* $\beta_k - \beta_{k-1} + \cdots \pm \beta_0 \leqslant 0$ *then* $A_k = 0$.

Conversely if the dimension n of M is $\geqslant 5$, M has torsion-free homology, and M is simply connected then to any set of nonnegative integers $\{A_k\}_{k=1}^{n-1}$ satisfying (a), (b) and (c) above, there corresponds a nonsingular Morse-Smale flow with A_k untwisted closed orbits of index k and no twisted closed orbits.

Note that property (a) is simply the Morse inequalities (8.6) of Smale and since $A_n = A_{n+1} = 0$ they imply that the Euler characteristic of M vanishes. For the proof of the remainder of the theorem we refer the reader to [F5].

Chapter 9. Smale flows

In this chapter we consider a more general class of flows than Morse-Smale, allowing basic sets which are one dimensional, but with infinitely many orbits.

(9.1) DEFINITION. If f_t is a flow on M it is called a *Smale flow* provided

(a) The chain recurrent set R of f_t has a hyperbolic structure.

(b) The set R is one dimensional.

(c) The flow satisfies the transversality condition.

Smale flows are the natural analogue for flows of Smale diffeomorphisms. To make this more precise we need the following definition.

(9.2) DEFINITION. If $h: X \longrightarrow X$ is a homeomorphism then its *suspension flow* (or *mapping torus flow*) is defined on the identification space $Y = X \times R/(x, s + 1) \sim (h(x), s)$, and is defined to be the flow on Y induced by the flow ϕ_t on $X \times R$ given by $\phi_t(x, s) = (x, t + s)$.

We list several elementary properties of suspension flows which are easy to check.

(9.3) PROPOSITION. *If $h: X \longrightarrow X$ is a homeomorphism and $f_t: Y \longrightarrow Y$, $t \in R$, is its suspension flow then*

(a) *If X is a manifold and h is a diffeomorphism then Y is a manifold and f_t is a smooth flow.*

(b) *If $h_1, h_2: X \longrightarrow X$ are isotopic diffeomorphisms then the corresponding suspension flow spaces Y_1 and Y_2 are diffeomorphic.*

(c) *There is a fibration $p: Y \longrightarrow S^1$, with the flow transverse to all fibers and each fiber homeomorphic to X. The map $p: Y \longrightarrow S^1$ is induced by the projection $X \times R \longrightarrow R$ if we consider $S^1 = R/s \sim (s + 1)$.*

(d) *If X is compact, taking suspensions defines a one-to-one correspondence between homeomorphisms $h: X \longrightarrow X$ and pairs (f_t, p) where $p: Z \longrightarrow S^1$ is a fibration with fiber X and f_t is a flow on Z transverse to all fibers.*

It is important to note that the suspension flow alone does not determine the map h; one also needs the fibration $p: Z \longrightarrow S^1$ as in (d) above. Two quite different diffeomorphisms can have the same suspension flow.

We want now to investigate the structure of a one dimensional basic set Λ for a flow f_t with hyperbolic chain recurrent set. Since Λ is one dimensional, if we choose a small disk D of dimension $(n - 1)$ which is transverse to the flow, the intersection of D and Λ is zero dimensional. This means that it is possible to choose a submanifold N_0 of D, also of dimension $(n-1)$, such that $\partial N_0 \cap \Lambda = \emptyset$. (However, it may not be possible to make N_0 a disk!) We form a transversal N made up of a number of components each like N_0. We choose enough that N crosses every orbit of Λ, and note that $N \cap \Lambda \subset \mathrm{int}\, N$ since each component of N had this property.

Now let $\Omega = N \cap \Lambda$. For a small neighborhood Q of Ω in N we can define the first return map $g: Q \rightarrow N$. To be precise we let $\tau(x) = \inf\{t > 0 \mid f_t(x) \in N\}$ and note that τ is continuous (in fact smooth) for x in some neighborhood Q of Ω in N. We then let $g: Q \rightarrow N$ be defined by $g(x) = f_{\tau(x)}(x)$.

The following lemma is now an easy exercise.

(9.4) LEMMA. *The compact invariant set Ω of $g: Q \rightarrow N$ has a hyperbolic structure and is in fact a basic set for g.*

It is true that we have only defined basic sets for invertible maps, but there is no reason to exclude embeddings like $g: Q \rightarrow N$. Choosing Q to be a compact manifold with boundary it is easy to show that Ω is the chain recurrent set of g. We have almost given the proof of a theorem of Bowen.

(9.5) THEOREM [B4]. *If Λ is a one dimensional basic set of a flow f_t with hyperbolic chain recurrent set then $f_t|\Lambda$ is topologically equivalent to the suspension of a subshift of finite type.*

PROOF. We first note that $f_t|\Lambda$ is topologically equivalent to the suspension flow of the first return map $g|\Omega: \Omega \rightarrow \Omega$. In fact we can consider the space of the suspension flow to be

$$Y = \Omega \times [0, 1]/(x, 1) \sim (g(x), 0)$$

and then the topological equivalence $\Psi: Y \rightarrow \Lambda$ is given by $\Psi((x, t)) = f_{t\tau(x)}(x),\ 0 \leqslant t < 1$.

Next we observe that since Ω is a zero dimensional basic set of g, the result of Bowen (3.14), or more precisely the special case which we prove in (A.5), says that $g|\Omega$ is topologically conjugate to a subshift of finite type. Q.E.D.

This result motivates the following definition.

(9.6) DEFINITION. Two nonnegative integral square matrices A and B are *flow equivalent* provided the subshifts of finite type $\sigma(A)$ and $\sigma(B)$ have topologically equivalent suspension flows. A characterization of this equivalence relation was given by Parry and Sullivan.

(9.7) THEOREM [PS]. *The equivalence relation of flow equivalence is generated by strong shift equivalence and the relation*

$$\begin{pmatrix} a_{11} & \cdots & a_{1n} \\ & \cdots & \\ a_{n1} & \cdots & a_{nn} \end{pmatrix} \sim \begin{pmatrix} 0 & 1 & 0 & \cdots & 0 \\ a_{11} & 0 & a_{12} & \cdots & a_{1n} \\ a_{21} & 0 & a_{22} & \cdots & a_{2n} \\ & & & \cdots & \\ a_{n1} & 0 & a_{n2} & \cdots & a_{nn} \end{pmatrix}.$$

This means that if A and B are flow equivalent there exist matrices $A_1 = A, A_2, \ldots,$ $A_m = B$ such that A_i is strong shift equivalent to A_{i+1} or A_i has the form of one of the two matrices above and A_{i+1} has the form of the other.

(9.8) COROLLARY [PS]. *The number* $\Psi(A) = \det(I - A)$, *called the Parry-Sullican invariant, is an invariant of flow equivalence.*

PROOF. We showed in Chapter 5 that even the function $\det(I - At)$ is an invariant of strong shift equivalence. Expansion by minors shows that $\Psi(A)$ is an invariant of the other relation. Q.E.D.

The study of a flow f_t on a one dimensional basic set Λ by studying the first return map $g: \Omega \longrightarrow \Omega$ on Λ intersected with a transversal N (see (9.4)) is an important technique which we will continue to use. All our properties and definitions for zero dimensional basic sets of diffeomorphisms can be applied. In particular we want to define the notion of structure matrix as in (4.10). Nothing is really different; we define for $x \in \Omega$

$$
\Delta(x) = \begin{cases} 1 & \text{if } D_{g_x}: E_x^u \longrightarrow E_{g(x)}^u \text{ preserves orientation,} \\ -1 & \text{otherwise,} \end{cases}
$$

recalling from (9.4) that Ω is a basic set for the first return map $g: Q \longrightarrow N$. The function $\Delta: \Omega \longrightarrow \{1, -1\}$ is continuous and hence locally constant. As before we choose a vertex shift $\sigma(B): \Sigma_B \longrightarrow \Sigma_B$ conjugate to $g|\Omega$, say by $h: \Sigma_B \longrightarrow \Omega$, and we choose it in such a way that Δ is constant on $h(\{\mathbf{b} \in \Sigma_b | b_0 = k\})$. Call the value of Δ on this set Δ_k.

(9.9) DEFINITION. The matrix $A = (A_{jk})$ where $A_{jk} = \Delta_k B_{jk}$ is called a *structure matrix* for Λ.

For example if $g: Q \longrightarrow N$ is hyperbolic with respect to a set of handles then perhaps by choosing a set of smaller but more numerous handles we can arrange that the geometric intersection matrix G contains only zeroes and ones. In this case B can be G and the structure matrix A is just the algebraic intersection matrix. As with diffeomorphisms it is useful to use a Lyapunov function to give a filtration for flows.

(9.10) DEFINITION. If $f_t: M \longrightarrow M$ is a flow with hyperbolic chain recurrent set and its basic sets are $\{\Lambda_i\}_{i=0}^n$ then a *filtration associated to* f_t is a collection of submanifolds $M_0 \subset M_1 \subset \cdots \subset M_n = M$ such that
   (a) $f_t(M_i) \subset \text{int } M_i$, for all $t > 0$,
   (b) $\Lambda_i = \bigcap_{t=-\infty}^{\infty} f_t(M_i - M_{i-1})$.
The existence of such a filtration is obtained exactly as in (4.8). The following result relates the structure matrix of the basic set Λ_i to the homology of the pair (M_i, M_{i-1}).

(9.11) THEOREM [BF]. *If f_t is a Smale flow with associated filtration $\{M_j\}$ and B is an $n \times n$ structure matrix for the basic set Λ_i of index u then*

$$H_k(M_i, M_{i-1}; F) = 0 \quad \text{if } k \neq u, u+1,$$

$$H_u(M_i, M_{i-1}; F) = F^n/(I - B)F^n,$$

$$H_{u+1}(M_i, M_{i-1}; F) = \ker((I - B) \text{ on } F^n)$$

for any abelian group F.

SKETCH OF PROOF. We consider only the case when F is a field. Choose Q and N as in (9.4) and so that $N \subset M_i - M_{i-1}$, and Q is a manifold with boundary. Define the pair (X, A) by $X = M_{i-1} \cup (\bigcup_{t>0} f_t(Q))$ and $A = \text{closure}(X - \{f_t(x) | x \in Q, 0 \leqslant t \leqslant \tau(x)\})$. Let $\hat{X} = A \cup Q$ and extend τ from Q to a positive continuous function $\tau: \hat{X} \rightarrow R$. Define $g: (\hat{X}, A) \rightarrow (\hat{X}, A)$ by $g(x) = f_{\tau(x)}(x)$ so g extends $g|Q$. The fact that $\Omega = \Lambda \cap N$ is a basic set for $g: Q \rightarrow N$ is sufficient to obtain the conclusion of Theorem (4.11) applied to the pair $(\hat{X}, A)$. In particular

(a) $g_*: H_k(\hat{X}, A; F) \rightarrow H_k(\hat{X}, A; F)$ is nilpotent if $k \neq u$.

(b) The nonnilpotent part of B is conjugate (similar) to the nonnilpotent part of g_{*u}.

From this it is an easy algebra exercise to show that for $g_*: H_u(\hat{X}, A; F) \rightarrow H_u(\hat{X}, A; F)$ and $B: F^n \rightarrow F^n$,

$$\ker(I - B) \cong \ker(I - g_*) \quad \text{and} \quad \text{coker}(I - B) \cong \text{coker}(I - g_*).$$

We next need a lemma.

(9.12) LEMMA. *If $X \supset \hat{X} \supset A$ are as above there is a long exact sequence*

$$\cdots \rightarrow H_{k+1}(X, A) \xrightarrow{\phi} H_k(\hat{X}, A) \xrightarrow{I - g_*} H_k(X, A) \rightarrow \cdots.$$

PROOF. We consider the commutative diagram

$$\cdots \rightarrow H_{k+1}(X, A) \xrightarrow{j_*} H_{k+1}(X, \hat{X}) \xrightarrow{\partial_1} H_k(\hat{X}, A) \rightarrow H_k(X, A) \rightarrow \cdots$$

$$h_* \big\uparrow \cong \qquad \qquad \overline{h}_* \big\uparrow \cong \quad \text{excision}$$

$$H_{k+1}((\hat{X}, A) \times (J, \partial J)) \xrightarrow{\partial_2} H_k((\hat{X} \times \partial J) \cup (A \times J), A \times J)$$

$$k \big\uparrow \cong \qquad \qquad i_* \big\uparrow \cong$$

$$H_k(\hat{X}, A) \xrightarrow{(I, -1)} H_k(\hat{X}, A) \oplus H_k(\hat{X}, A)$$

where the top row is the long exact sequence for the triple $A \subset \hat{X} \subset X$, and h_* is induced by the map $h: (\hat{X}, A) \times (J, \partial J) \rightarrow (X, \hat{X})$ given by $h(x, t) = f_{t\tau(x)}(x)$ ($t \in J = [0, 1]$). The map h is a relative homeomorphism so h_* is an isomorphism; $\overline{h}$ is the restriction of h. The map k is an isomorphism coming from the Kunneth formula and ∂_2 is the boundary map of the long exact sequence of the triple $(A \times J) \subset (\hat{X} \times \partial J) \cup (A \times J) \subset (\hat{X} \times J)$. The map $i: (\hat{X}, A) \amalg (\hat{X}, A) \rightarrow ((\hat{X} \times \partial J) \cup (A \times J), A \times J)$ takes the first copy of $(\hat{X}, A)$ to $(\hat{X}, A) \times \{0\}$ and the second to $(\hat{X}, A) \times \{1\}$.

One checks easily that $h_* \circ i_*(u, v) = u + g_*(v)$, so that $h_* \circ i_* \circ (I, -I) = I - g_*$. Thus tracing through the diagram shows the required sequence is exact when $\phi = k^{-1} \circ h_*^{-1} \circ j_*$. Q.E.D.

We return to the proof of (9.11). For $k \neq u$, g_* is nilpotent so $(I - g_*)$ is an isomorphism; hence the exact sequence above allows one to conclude $H_k(X, A) = 0$ if $k \neq u$, $u + 1$. On the other hand the sequence

$$0 \longrightarrow H_{u+1}(X, A) \longrightarrow H_u(\hat{X}, A) \xrightarrow{I - g_*} H_u(\hat{X}, A) \longrightarrow H_u(X, A) \longrightarrow 0$$

is exact so

$$H_{u+1}(X, A) \cong \ker(I - g_*) \cong \ker(I - B)$$

and

$$H_u(X, A) \cong \operatorname{coker}(I - g_*) \cong \operatorname{coker}(I - B).$$

To conclude the proof we need only observe that if T is sufficiently large $f_T \colon (X, A) \to (M_i, M_{i-1})$ and $f_T \colon (M_i, M_{i-1}) \to (X, A)$ are defined and since f_{2T} is homotopic to the identity the pairs (X, A) and (M_i, M_{i-1}) are homotopy equivalent. Q.E.D.

REMARK. Note that if $\det(I - B) \neq 0$ then $|\det(I - B)| = $ order of $\operatorname{coker}(I - B) = $ order of $H_u(M_i, M_{i-1}; Z)$. Compare this with the Parry-Sullivan invariant (9.7).

We now proceed to some Morse inequalities for Smale diffeomorphisms due to Zeeman.

(9.13) THEOREM [Z]. *Suppose f_t is a Smale flow on M and c_k is the number of rest points of index k. For each one dimensional basic set $\Lambda_i(k)$ of index k choose a structure matrix $A_i(k)$ and define $v_k(F) = \Sigma_i \dim \ker(I - A_i(k))$ where we consider $(I - A_i(k))$: $F^n \to F^n$, with F a field and n the size of $A_i(k)$. Then*

$$v_k(F) + c_k - c_{k-1} + \cdots \pm c_0 \geqslant \beta_k(F) - \beta_{k-1}(F) + \cdots \pm \beta_0(F)$$

where $\beta_i(F) = \dim H_i(M; F)$.

PROOF. The proof we give is for manifolds M with the flow entering on the boundary, and proceeds by induction on the number of basic sets. If the flow has only one basic set, it must have index 0 by (1.13). Hence by (1.15) it must be a rest point or a single closed orbit. Thus using (9.11) above the theorem is easy if there is only one basic set. We now assume it true for all flows with fewer than n basic sets (and for all k). If f_t on M has n basic sets we choose an associated filtration $M_0 \subset M_1 \subset \cdots \subset M_n$. By the induction hypothesis the result holds for f_t restricted to M_{n-1}.

Suppose the basic set Λ_n is not a rest point. We consider the exact sequence of the pair (M_n, M_{n-1}) and the fact from (9.11) that

$$H_k(M_n, M_{n-1}; F) \cong \begin{cases} F^r & \text{if } k = u \text{ or } u - 1, \\ 0 & \text{otherwise} \end{cases}$$

where $u = \text{index } \Lambda_n$ and $r = \dim \ker(I - A_n)$. Letting $X = M_n$ and $A = M_{n-1}$, the sequence becomes

$$\cdots \longrightarrow 0 \longrightarrow H_{u+1}(A) \longrightarrow H_{u+1}(X) \longrightarrow F^r \overset{\alpha}{\longrightarrow} H_u(A) \longrightarrow H_u(X) \longrightarrow F^r \overset{\beta}{\longrightarrow}$$

$$H_{u-1}(A) \longrightarrow H_{u-1}(X) \longrightarrow 0 \longrightarrow H_{u-2}(A) \longrightarrow H_{u-2}(X) \longrightarrow 0 \longrightarrow \cdots .$$

Define $B_k(X, A) = (-1)^k \Sigma_{i=0}^k \dim H_k(X, A)$ and define $B_k(X)$ and $B_k(A)$ similarly. Considering the exact sequence above as a chain complex with trivial homology we can apply (2.14) and obtain $B_k(X) = B_k(A)$ for all $k \neq u, u - 1$. For the case $k = u$ we extract the exact sequence

$$0 \longrightarrow \ker(\alpha) \longrightarrow H_u(A) \longrightarrow H_u(X) \longrightarrow F^r \longrightarrow H_{u-1}(A) \longrightarrow \cdots$$

and conclude $B_u(X) = B_u(A) + r - \dim \ker(\alpha)$ so $B_u(X) \leq B_u(A) + r$. Similarly $B_{u-1}(X) = B_{u-1}(A) - \dim \ker \beta$ so $B_{u-1}(X) \leq B_{u-1}(A)$. By the induction hypothesis

$$v_k + (c_k - c_{k-1} + \cdots \pm c_0) \geq B_k(A)$$

where v_k is defined for the flow $f_t | A$. We want to show $v_k' + (c_k - c_{k-1} + \cdots \pm c_0) \geq B_k(X)$ where v_k' is defined for f_t on X. However since

$$v_k' = \begin{cases} v_k & \text{if } k \neq u, \\ v_u + r & \text{if } k = u \end{cases}$$

this result clearly follows from the inequalities above. The case when Λ is a rest point is treated similarly. We leave the details to the reader.   Q.E.D.

There have been a number of attempts to develop a zeta function theory for flows (see [S1]). Since for most of these the function associated to a flow depends on parametrization it is not even an invariant of topological equivalence, and thus cannot be expected to be related to homological invariants. We will not discuss this treatment of zeta functions for flows but instead consider a more homological approach which has the serious drawback that it gives a nontrivial invariant only for manifolds M with $H_1(M) \neq 0$.

(9.14) DEFINITION [Fr1]. A flow f_t is *homology finite* provided there are finitely many closed orbits (distinguishing multiples) in each homology class in $H_1(M; Q)$.

In particular there can be no periodic orbit which is null homologous since the same orbit with multiple periods would form an infinite set all representing the 0 homology class.

The following definition is due to Williams and Fried (see [Fr2, Wm3]).

(9.15) DEFINITION. If f_t is a homology finite flow on M and $H = H_1(M)/\text{torsion}$ then $\zeta(f_t)$ is defined by

$$\log \zeta(f_t) = \sum_\gamma \sum_{k=1}^\infty \frac{[\gamma]^k}{k}$$

where the first sum is over all least period closed orbits γ and $[\gamma] \in Z[H]$ is the element determined by γ (H is written multiplicatively here). Thus $\log \zeta(f_t)$ is in $Q[[H]]$, the formal power series in elements of H with coefficients in Q.

The main result concerning this zeta function is the following rationality theorem of Fried.

(9.16) THEOREM [**Fr2**]. *If f_t is homology finite and has a hyperbolic chain recurrent set then $\zeta(f_t)$ can be expressed in the form $(1 + P)/(1 + Q)$ with $P, Q \in Z[H]$.*

One can also form a homology zeta function for homology finite flows. We have the following result of Fried.

(9.17) THEOREM [**Fr2**]. *Suppose f_t is a homology finite Smale flow and*

$$Z(f_t) = \prod_\gamma (1 - \Delta_\gamma [\gamma])^{(-1)^{u(\gamma)}}$$

where $\Delta_\gamma = -1$ if γ is a twisted closed orbit and 1 otherwise and $u(\gamma)$ is the index of γ. Then $Z(f_t) \in Q[[H]]$ is a Reidemeister torsion invariant of M, and in particular is independent of the flow f_t.

We will not give the proof. However this result grew out of and generalizes many cases of (10.5), which we discuss in the next chapter. The result (10.5) is concerned with knot and link complements only but does not require the hypothesis of homology finiteness.

Chapter 10. Flows on S^3

In our earlier discussion of diffeomorphisms it became evident that there is a dichotomy between the study of systems on high dimensional simply connected manifolds and systems on low dimensional manifolds. This also holds true for flows.

While the study of diffeomorphisms on surfaces yields a rich and apparently complex theory, the study of flows on surfaces with hyperbolic chain recurrent set is fairly well understood through the work of Peixoto [P]. It is the investigation of flows on three dimensional manifolds which leads to interesting and deep results.

There is now such a variety of good results about flows in three dimensions that it is impossible to survey them all here. In particular the two papers of Birman and Williams [BW] on knotted periodic orbits are to be recommended.

Our remarks here will be directed toward flows on S^3 with particular attention to results with a homological flavor.

The following proposition is closely related to an example given by F. B. Fuller and can be found in Asimov [As3].

(10.1) PROPOSITION. *There exists a nonsingular Morse-Smale flow on the solid torus $S^1 \times D^2$ which is inwardly transverse to the boundary and has chain recurrent set consisting of two untwisted, unlinked closed orbits, one of index zero and one of index one. Both closed orbits are null homotopic.*

SKETCH OF PROOF. Consider the vector field on the two dimensional region sketched in Figure (10.1) which has two rest points, a sink and a saddle. If this region is rotated around the y-axis and the vector field carried by the same rotation we obtain a vector field on a three dimensional solid with two circles of rest points corresponding to the two rest points in the x-y plane. A slight perturbation supported on a neighborhood of these circles will change them to closed orbits of index 0 and 1. Finally the top of the "bottle" we have formed is glued to the bottom in such a way that all trajectories passing through the glued region tend toward the index zero closed orbit, i.e. toward the sink.

The resulting flow is Morse-Smale and inwardly transverse to the boundary (which can be rounded off at its corners). Q.E.D.

(10.2) THEOREM [F4]. *Necessary and sufficient conditions for the existence of a nonsingular Morse-Smale flow on S^3 with A_k untwisted closed orbits of index k are*

(a) $A_0 \geqslant 1, A_2 \geqslant 1.$

(b) $A_1 \geqslant A_0 - 1, A_1 \geqslant A_2 - 1.$

With any specified numbers of untwisted orbits the number of twisted orbits of index 1 is completely arbitrary. There can be no twisted orbits of index 0 or 2.

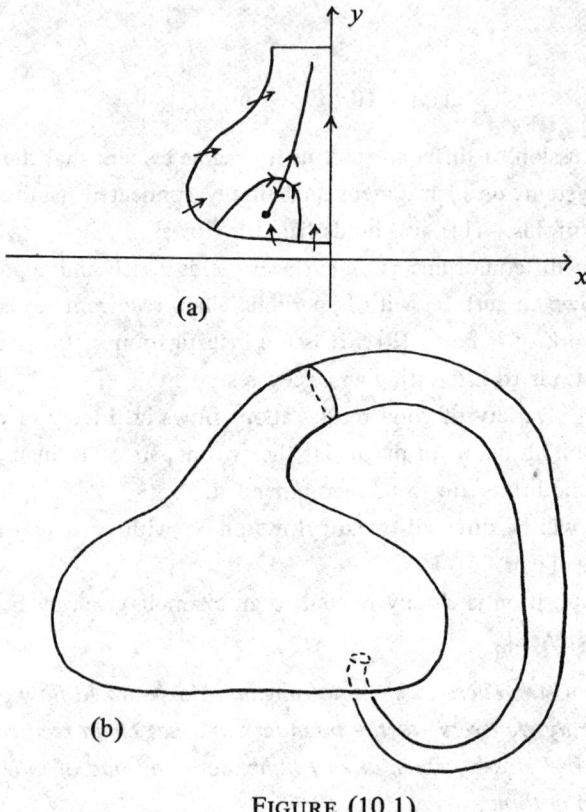

FIGURE (10.1)

PROOF. We first note that the necessity of (a) and (b) follows from (a) and (b) of Theorem (8.9).

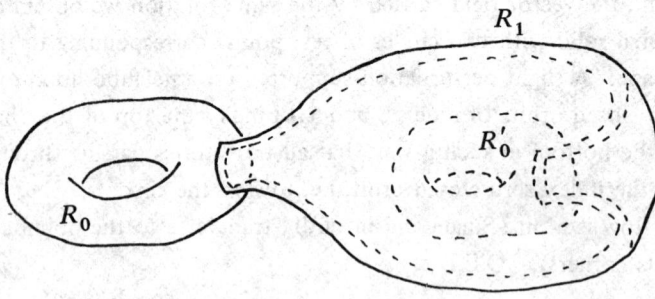

FIGURE (10.2)

For the sufficiency we first consider Figure (10.2). There are two solid tori $(S^1 \times D^2)$ labelled R_0 and R_0' on which we can put a vector field inwardly transverse to the boundaries and with an attracting (index zero) closed orbit being the only chain recurrent points. To R_0 and R_0' we attach $R_1 = S^1 \times D^1 \times D^1$. It is attached as shown by identifying $S^1 \times D^1 \times S^0$ with embedded copies of $S^1 \times D^1$ in the boundary of R_0 and R_0'. On R_1 we can construct a vector field with a single untwisted closed orbit of index 1 in such a way that the vector field agrees with the vector field on R_0 and R_0' where identifications are made and is transverse inward on the remainder of R_1. In coordinates (θ, x, y) for $R_1 = S^1 \times D^1 \times D^1$ this vector field could be $X = \partial/\partial\theta - x(\partial/\partial x) + y(\partial/\partial y)$.

Noting now that the complement of $R_0 \cup R_0' \cup R_1$ in S^3 is two disjoint copies of $S^1 \times D^2$, we add two such solid tori on each of which an outward pointing vector field with one closed orbit of index 2 has been constructed. These vector fields are constructed to match up with those already defined on the boundary of $R_0 \cup R_0' \cup R_1$. We have thus constructed a Morse-Smale flow on S^3 with $A_0 = 2, A_1 = 1, A_2 = 2$.

We can however iterate this construction as shown in Figure (10.3) to create a flow with $A_0 = m, A_1 = m - 1, A_2 = m$.

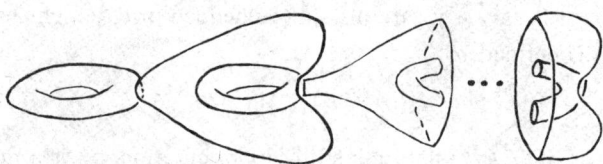

FIGURE (10.3)

For the general case we suppose that A_0, A_1, A_2 satisfying (a) and (b) are given and let $m = \min\{A_0, A_2\}$. We consider the case $A_2 = m$ the other being similar. First, as above we can construct a flow with m closed orbits of index 0, $m - 1$ of index 1, and m of index 2.

For embeddings $D^2 \rightarrow D^2$ it is easy by a local change to replace a single sink with two sinks and a saddle. Taking suspensions allows one to replace a closed orbit of index zero by two of index 0 and one of index 1 (untwisted). Repeatedly using this fact we add $A_0 - m$ untwisted closed orbits of index 0 and the same number of index 1. Finally we use Proposition (10.1) to replace a tubular neighborhood of an index 0 closed orbit with a copy of $S^1 \times D^2$ possessing a flow with one orbit of index 0 and one of index one. Repeating this if necessary we can add $A_1 - A_0$ untwisted orbits of index 1.

It is clear that there can be no twisted orbits of index 0 or 2 since it is not possible to embed an unoriented three-manifold in S^3.

To see that the number of twisted orbits of index 1 is arbitrary we consider the following example. It is not difficult to construct an embedding of the disk D^2 in its interior with precisely three hyperbolic periodic points: a sink of period 2 and a saddle whose unstable manifold has its orientation reversed. By taking the suspension (or mapping torus) of this embedding and rounding off corners we obtain a flow on the solid torus pointing inward on the boundary and with one untwisted orbit of index zero and one twisted orbit of index one.

If we now take any nonsingular Morse-Smale flow on S^3, cut out a tubular neighborhood of an index zero closed orbit and replace it by the example above we have increased the number of twisted closed orbits by one without changing the number of untwisted ones. Repeated application will give any desired number of twisted closed orbits of index one. Finally we take a Kupka-Smale approximation (see [S1]) to achieve transversality of stable and unstable manifolds, and this will not change the periodic behavior.   Q.E.D.

It is clear that the replacement constructions used in the proof above drastically change the way closed orbits link one another and properties like the knot type of closed orbits. Both of these are natural things to include in a qualitative description of a flow. We proceed to do that now, but in the more general context of Smale diffeomorphisms, which of course includes the Morse-Smale systems we have been considering.

If γ is a smoothly embedded oriented circle in S^3 and γ' is a *meridian* of a tubular neighborhood N of γ, i.e., a circle embedded in ∂N which is contractible in N, then the orientations of S^3 and γ determine an orientation of γ', say by the right-hand rule. It is a consequence of the Alexander duality theorem that $H_1(S^3 - \gamma) \cong Z$ and $[\gamma']$ is a generator.

(10.3) DEFINITION. If γ_1, γ_2 are disjoint embedded oriented circles in S^3 then their *linking number* $l(1, 2)$ is defined by

$$i_*(\alpha) = l(1, 2)[\gamma_2']$$

where α is the generator of $H_1(\gamma_1)$ determined by its orientation, γ_2' is a meridian of γ_2 and $i_*: H_1(\gamma_1) \rightarrow H_1(S^3 - \gamma_2)$ is induced by inclusion.

Note that changing the orientation of either γ_1 or γ_2 changes the sign of $l(1, 2)$. We will adopt the convention of always orienting a closed orbit of a flow by the direction of the flow.

If f_t is a nonsingular Smale flow on S^3 we can include even more information in the structure matrices.

In (9.4) we considered a one dimensional basic set Λ and constructed a transversal N to the flow, crossing every orbit of Λ. Recall from (9.5) that if $\Omega = N \cap \Lambda$ and $g: \Omega \rightarrow \Omega$ is the first return map then g is topologically conjugate, say by h, to a subshift of finite type $\sigma(A): \Sigma_A \rightarrow \Sigma_A$. Define the sets $C_j = h(\{a \in \Sigma_A | a_0 = j\})$ and note that each orbit of Λ is uniquely determined by the sequence of C_j's it passes through. As in the definition of structure matrix (9.9) we want to choose the shift $\sigma(A)$ in such a way that the C_j's are so small that the function Δ is constant on each of them. We wish now to describe how orbits going from C_j to C_k link a given knot (or link) which is a closed orbit attractor or repeller (or set of such) for the flow f_t.

If the C_j's above are chosen with sufficiently small diameter then it is possible to choose a smooth disk D_j transverse to the flow which contains C_j and with $D_j \subset N$. (In doing this we may alter C_j slightly, moving points of C_j along orbits of the flow a short distance so they lie in D_j.)

Let L be a designated set of closed orbits which are attractors or repellers for the flow f_t. We choose a base point in $S^3 - L$ and paths from it to each of the D_j. We assume that D_j was chosen so $D_j \cap L = \varnothing$. Now any curve from C_j to C_k can be closed by means of the curves from the base point to D_j and D_k and we can thus define its linking number (using the flow direction for orientation) with each component of L.

If the C_j have sufficiently small diameter, then all orbit segments going from C_j to C_k (and intersecting no other C in between) will have the same linking number with a given component of L. We assume the C's are small enough that this is the case for all components of L and then we may speak simply of the linking number of a component of L with all the orbit segments from C_j to C_k.

(10.4) DEFINITION. Suppose f_t is a nonsingular Smale flow with a basic set Ω which has B as a structure matrix, and L is an oriented link consisting of closed orbits of the flow of index 0 or 1. The matrix $S = (s_{ij})$ defined by $s_{ij} = b_{ij} t_1^{n_1} t_2^{n_2} \cdots t_n^{n_k}$, where n_p is the linking number of the orbit segments from C_i to C_j with the pth component of L, will be called a *linking matrix* for Ω with respect to L.

The entries of S are polynomials in the indeterminates $\{t_p\}$ and their inverses. Of course any basic set will have many linking matrices: there were choices for B as well as for base point and paths to the D_j's. We will see however that $\det(I - S)$ is unique up to multiples of $\pm t_p^{\pm 1}$.

Our aim is to investigate Smale flows on S^3 and how the existence of a knot (or link) as repeller (or repellers) in the flow affects the dynamics of the remainder of the flow. We will do this by considering the flow on a manifold M (with boundary) which is obtained by deleting from S^3 tubular neighborhoods of the components of the link L in question. For the moment, we will assume each component of L in S^3 is a repelling closed orbit of the flow (there may be additional such closed orbit repellers which are not in L). Thus the flow on M comes from a vector field which is inwardly transverse to the boundary.

It follows from Alexander duality that $H_1(M) = Z^k$ where k is the number of components of the link L. An important tool in studying L or M is a certain covering space of M called the *universal abelian cover* which we now describe. Let $G = \ker(\Pi_1(M) \to H_1(M))$; then G is a normal subgroup of $\Pi_1(M)$ and we let $\widetilde{M} \xrightarrow{P} M$ denote the covering with group G (i.e., $\Pi_1(\widetilde{M}) = G$ and the group of covering transformations is $\Pi_1(M)/G \cong H_1(M) \cong Z^k$). We can obtain a cell decomposition of $\widetilde{M}$ by choosing a finite triangulation of M (which is compact) and lifting it to $\widetilde{M}$. The covering transformations then act freely on the simplices of $\widetilde{M}$ and we can consider the chain groups for $\widetilde{M}$ as finitely generated free modules over the group ring of the group of covering transformations $Z(H_1(M))$. More precisely, we choose a distinguished basis $t_1, t_2, \ldots, t_k$ of $H_1(M)$ corresponding to oriented meridians of the components of the link L. Then each t_i corresponds to a covering transformation of $\widetilde{M}$ (since the covering transformations are $\Pi_1(M)/G = H_1(M)$). These covering transformations are closely related to the way closed curves in M link the components of L. For example, if a closed curve links L so the linking number with the ith component is n_i and the curve is lifted to an arc $\gamma: [0, 1] \to \widetilde{M}$ in $\widetilde{M}$, then we will have $\gamma(1) = t_1^{n_1} t_2^{n_2} \cdots t_k^{n_k}(\gamma(0))$.

If σ is a simplex in $\widetilde{M}$, $t_i(\sigma)$ is another simplex which we think of as t_i times σ, and the chains in $\widetilde{M}$ can thus be made into a module over the ring $K = Z[t_1, t_1^{-1}, t_2, t_2^{-1}, \ldots, t_k, t_k^{-1}]$

of integer polynomials in the t_i's and their inverses (this is $Z(H_1(M))$, with $H_1(M)$ written multiplicatively).

We can then of course calculate the homology using the K module structure on these chains and obtain homology groups which we denote $H_*(\widetilde{M})$.

If $M_1 \subset M_2$ are subcomplexes of M, we can form $\widetilde{M}_i = P^{-1}(M_i)$ and similarly consider $H_*(\widetilde{M}_2, \widetilde{M}_1)$. Of course, the homology groups $H_*(\widetilde{M}_i)$ and $H_*(\widetilde{M}_2, \widetilde{M}_1)$ are also K modules.

To any finitely generated K module P one can associate an element $A(P) \in K$ well defined up to multiples of $t_j^{\pm 1}$, $1 \leqslant j \leqslant k$, which is called the Alexander polynomial, and the *Alexander polynomial* Δ_L of a link L is defined to be $A(H_1(\widetilde{M}))$ where M is the complement of a tubular neighborhood of L. The precise definition of $A(P)$ can be found in [F6].

(10.5) THEOREM [F6]. *Suppose f_t is a nonsingular Smale flow on S^3, L is a link consisting of n closed orbits oriented by the flow, each an attractor or repeller, and $\{S_i\}$ are the linking matrices of the basic sets of index one, with respect to L. Then if $n > 1$,*

$$\prod_i \det(I - S_i) = \Delta_L(t_1, \ldots, t_n)\prod_k (1 - t_1^{l(1,k)} \cdots t_n^{l(n,k)})$$

up to multiples of $\pm t_j^{\pm 1}$, $1 \leqslant j \leqslant n$, where $l(j, k)$ is the linking number of the jth component of L with the kth component of the set of attractors and repellers not in L.

If L is a knot, i.e., $n = 1$, then

$$(1 - t) \prod_i \det(I - S_i) = \Delta_L(t)\prod_k (1 - t^{l(1,k)})$$

up to multiples of $\pm t^{\pm 1}$.

The equalities in this theorem are valid only modulo multiples of $\pm t_j^{\pm 1}$ since the Alexander polynomial is only defined up to such multiples. We remark also that any vacuous products in the theorem above are taken to be 1. Thus if L consists of all attractors and repellers of f_t, we have (modulo $\pm t_j^{\pm 1}$, $1 \leqslant j \leqslant n$) $\Delta_L(t_1, \ldots, t_n) = \Pi_i \det(I - S_i)$.

In [F6] some classical results of Seifert and Torres on the Alexander polynomial of links are derived from the theorem above. For example, simply applying the theorem to the backwards flow f_{-t} shows that if L' is the link L with all orientations reversed then $\Delta_L(t_1, \ldots, t_n) = \Delta_{L'}(t_1, \ldots, t_n)$, which is in turn equal to $\Delta_L(t_1^{-1}, \ldots, t_n^{-1})$ modulo $\pm t_j^{\pm 1}$.

(10.6) DEFINITION. If B_i is a structure matrix for the basic set Λ_i, define the *Parry-Sullivan invariant* Ψ_i of Λ_i by

$$\Psi_i = |\det(I - B_i)|.$$

Note that the remark after (9.11) implies that Ψ_i depends only on the flow not on the choice of structure matrix B_i.

(10.7) COROLLARY [F7]. *If f_t is a nonsingular Smale flow on S^3 with a single attractor γ_s and a single repeller γ_u then $|l(\gamma_s, \gamma_u)| = \Pi_i \Psi_i$ where $l(\gamma_s, \gamma_u)$ is the linking number and the product $\Pi \Psi_i$ is over all basic sets Λ_i except γ_u and γ_s.*

PROOF. From (10.5) we have

$$\prod \det(I - S_i(t_1, t_2)) = \Delta_L(t_1, t_2)$$

where L is the link made up of γ_u and γ_s. Since $S_i(1, 1) = B_i$ is a structure matrix for Λ_i, we have $\Delta_L(1, 1) = \prod_i \det(I - B_i)$. It is a result of G. Torres [T] that $\Delta_L(1, 1) = l(\gamma_u, \gamma_s)$ so this completes the proof.   Q.E.D.

(10.8) COROLLARY [F7]. *If f_t is a nonsingular Morse-Smale flow on S^3 with one attractor γ_s and one repeller γ_u then the linking number*

$$l(\gamma_s, \gamma_u) = 0 \ or \ 2^k \quad for \ some \ integer \ k \geqslant 0.$$

If $l(\gamma_s, \gamma_u) = 2^k$ then there are exactly k twisted and no untwisted closed orbits of index 1.

PROOF. A structure matrix for a closed orbit is the one-by-one matrix (± 1) with $+$ for untwisted and $-$ for twisted closed orbits. Hence if we apply (10.7),

$$\Psi_i = \begin{cases} 0 & \text{if } \gamma_i \text{ is untwisted,} \\ 2 & \text{if } \gamma_i \text{ is twisted} \end{cases}$$

so

$$l(\gamma_s, \gamma_u) = \prod \Psi_i = 0 \ or \ \pm 2^k$$

Clearly if $l(\gamma_s, \gamma_u) = \pm 2^k$ there are k twisted and no untwisted orbits of index 1.   Q.E.D.

(10.9) COROLLARY. *If γ is a knotted closed orbit of a Morse-Smale flow on S^3 then the Alexander polynomial $\Delta_\gamma(t)$ has only roots which are roots of unity.*

PROOF. If γ has index one then f_t can be altered on a neighborhood of γ to introduce two new orbits, one of index 1 and one of index 2, both of the same knot type as γ (replace a saddle by two saddles and a source). The new flow will still be Morse-Smale. Hence we can assume without loss of generality that γ has index 0 or 2. Now applying (10.5) and using the fact that if Λ_i is a closed orbit $S_i = \pm t^{l(i)}$ where $l(i)$ is the linking number of γ with Λ_i, we have

$$(1 - t) \prod_i (1 \pm t^{l(i)}) = \Delta_\gamma(t) \prod_k (1 - t^{l(1,k)})$$

which gives the desired result.   Q.E.D.

A stronger result than this was proved by Morgan [Mo] who showed that a knotted closed orbit of a Morse-Smale flow on S^3 is in the class of knots generated from torus knots using the operations of cabling and connected sum.

Chapter 11. Beyond the theme

As we stated in the preface the theme of these lectures has been the relationship between dynamical properties of a flow or diffeomorphism and its homological configuration. The diagram

$$\underline{\text{Dynamics}} \xleftarrow{\quad\text{geometry}\quad} \begin{matrix}\text{Chain complex}\\ \text{description}\end{matrix} \xrightarrow{\quad\text{algebra}\quad} \underline{\text{Topology}}$$

(basic sets
and unstable
manifolds)

(Homological
configuration)

illustrates the connections between the two and the important intermediary role of chain complexes.

In this chapter we want to delve more deeply into the dynamics and relate the sharper dynamical picture to finer topological invariants of the manifold. We do this for the simplest dynamical systems—the gradient-like flows we considered in Chapter 2. The new element in the dynamical picture is a description of trajectories which run from one critical point p to another q, i.e., orbits asymptotic to q in backward time and p in forward time. The corresponding new elements in the topological configuration of the manifold are homotopy classes of attaching maps in the cell complex structure on M formed by the unstable manifolds. The intermediate role in this setting is played by the Pontryagin-Thom correspondence between homotopy classes of maps between spheres and framed submanifolds of spheres. All of this is explained in more detail below but we note the theme of this chapter, illustrated in the diagram,

$$\underline{\text{Dynamics}} \longleftrightarrow \begin{matrix}\text{Pontryagin-Thom}\\ \text{correspondence}\end{matrix} \longleftrightarrow \underline{\text{Topology}}$$

(connecting
manifolds)

(homotopy classes
of attaching maps)

closely parallels theme of earlier chapters, diagrammed above.

Suppose now that f_t is a flow on M which is gradient-like with respect to the Morse function $g: M \rightarrow R$, so the chain recurrent set R consists of the critical points of g which are rest points for f_t. The unstable manifolds $W^u(p), p \in R$, give a decomposition of M into cells.

We are interested in investigating the trajectories of f_t which "run from" one point p in R to another q. The situation is much simpler if there are no "intermediate" points of R between p and q.

(11.1) DEFINITION. If p, q are rest points of a gradient-like flow f_t we will say q is a *successor* to p provided $W^u(p) \cap W^s(q) \neq \varnothing$ but there is no rest point x with $W^u(p) \cap W^s(x) \neq \varnothing$ and $W^u(x) \cap W^s(q) \neq \varnothing$. The points p and q will be called *successive*.

Alternatively we could have defined successive by the requirement that closure($W^u(p)$) $\cap$ closure($W^s(q)$) $\cap R = \{p, q\}$.

(11.2) PROPOSITION. *If q is a successor to p then the set of trajectories running from p to q, i.e., the trajectories in $W^u(p) \cap W^s(q)$, form a manifold.*

PROOF. The fact that p and q are successive together with (1.14) says we can choose a Lyapunov function $\rho: M \to R$ such that $\rho(q) = a' < b' = \rho(p)$ and there are no other rest points in $\rho^{-1}([a', b'])$. It is an easy exercise to show that there are arbitrarily small $\epsilon > 0$ for which the function $h = \rho + \epsilon g$ is a Morse function with respect to which f_t is gradient-like. If ϵ is sufficiently small and $h(p) = b$, $h(q) = a$ then $a < b$ and $h^{-1}([a, b])$ contains no rest points other than p and q.

Now choose $c \in (a, b)$ and let $V = h^{-1}(c)$. Then in (2.4) we showed that the flow defines a diffeomorphism between $h^{-1}(b - \epsilon)$ and V. Also from (2.5) it is clear that $W^u(p) \cap h^{-1}(b - \epsilon)$ is a sphere of dimension (index of p) $- 1$. Hence $W^u(\rho) \cap V$ which we denote $S^u(p)$ is a sphere of this same dimension. Similarly $S^s(q) = W^s(q) \cap V$ is a sphere and the fact that $W^u(p)$ intersects $W^s(q)$ transversely in M implies that $S^u(p)$ and $S^s(q)$ intersect transversely in V. Thus the intersection $N = S^u(p) \cap S^s(q)$ is a manifold. Clearly each point x of N corresponds to a trajectory of the flow (namely the trajectory through x) which runs from p to q. Q.E.D.

(11.3) DEFINITION. The manifold $N = S^u(p) \cap S^s(q)$ is called a *connecting manifold* from p to q.

The first question we would like to address is, what kinds of manifolds can be connecting manifolds for some flow?

(11.4) DEFINITION. If N is an n dimensional submanifold of M^m then a *framing* of N in M^m is a continuous $(m - n)$-frame field for the normal bundle of N; or equivalently an ordered set of $(m - n)$ mutually orthogonal vector fields defined on N and orthogonal to N.

(11.5) PROPOSITION. *If q is a successor to p for the gradient-like flow f_t and N is a connecting manifold from p to q, then N has a framing in the sphere $S^u(p)$.*

PROOF. The plane $W^s(q)$ is contractible so it possesses a framing in M which is unique up to homotopy. As above let $V = h^{-1}(c)$. Then by (2.4) V has a neighborhood U in M diffeomorphic to $V \times [-1, 1]$ with $V \subset M$ corresponding to $V \times \{0\}$. Also $U \cap W^u(p)$ and $U \cap W^s(q)$ correspond to $S^u(p) \times [-1, 1]$ and $S^s(q) \times [-1, 1]$ respectively. From this it is clear that we can choose a Riemannian metric so that if v_x is a normal vector to $W^s(q)$ based at a point $x \in N$, then v_x is tangent to $S^u(p)$. It then is the case that the framing of $W^s(q)$ in M when restricted to N is a framing of N in $S^u(p)$. Q.E.D.

Manifolds which can be embedded in some sphere in such a way that they admit a framing are called *stably parallelizable*. The proposition above shows that any connecting manifold must be stably parallelizable. This excludes many manifolds, such as complex projective spaces CP^n, $n > 1$, from being connecting manifolds. We will see below in (11.10) that any stably parallelizable manifold can be a connecting manifold for some flow.

We now want to describe briefly the Pontryagin-Thom correspondence between framed cobordism classes of n dimensional manifolds in S^m and homotopy classes of maps from S^m to S^{m-n}. A good reference for this is [M4].

Two framed manifolds N_1 and N_2 in S^m are *cobordant* provided there is a manifold W^{n+1} framed in $S^m \times I$ such that $W \cap S^m \times \{0\}$ is the framed manifold $N_1 \subset S^m$ and $W \cap S^m \times \{1\}$ is the framed manifold $N_2 \subset S^m$. This defines an equivalence relation on n dimensional framed submanifolds of S^m. The correspondence between equivalence classes of framed manifolds and homotopy classes of maps from S^m to S^{m-n} is as follows:

(1) Given $f: S^m \longrightarrow S^{m-n}$ which is smooth, pick a regular value z and let $N = f^{-1}(z)$. The normal bundle to N is the pull-back $f^*(TS_z^{m-n})$. Pulling back a basis of TS_z^{m-n} gives a framing of this normal bundle.

(2) Conversely, given $N \subset S^m$ with framing define $f: S^m \longrightarrow S^{m-n}$ as follows. Let U be a tubular neighborhood of N in S^m. The framing provides coordinates on the fibers of U. Let $f(N) =$ north pole, $f(S^m - U) =$ south pole and on fibers of U, f maps rays to longitudes.

We want next to explore how the homotopy class of a connecting manifold is related to the topology of the manifold M. To do this we need to examine more carefully the cell structure provided to M by the unstable manifolds of the gradient-like flow f_t. We want to get a CW complex.

In associating a CW complex to a gradient-like flow we need an equivalence relation stronger than homotopy equivalence and weaker than homeomorphism, up to which the CW complex will be defined. The idea is that two CW complexes should be equivalent if they can inductively be built up by attaching corresponding cells with homotopic attaching maps.

If e and e' are cells of a CW complex Y, we will say that $e' \leqslant e$ if the closure of e contains any part of the interior of e'. If we make this relation transitive (and denote the new transitive relative by $\leqslant$ also) then we obtain a partial ordering on the cells of Y. If S is a subset of the cells of Y with the property that $e \in S$ and $e' \leqslant e$ implies $e' \in S$ then the union of the cells in S forms a subcomplex of Y. In particular if e is a cell of Y we define the *base* of e, denoted $Y(e)$, to be the smallest subcomplex of Y containing e. Thus $Y(e)$ is the union of all cells e' in Y such that $e' \leqslant e$.

(11.6) DEFINITION. Two finite CW complexes Y and Y' will be called *cell equivalent* providing there is a homotopy equivalence $h: Y \longrightarrow Y'$ with the property that there is a one-to-one correspondence between cells of Y and cells of Y' such that if $e \subset Y$ corresponds to $e' \subset Y'$ then h maps $Y(e)$ to $Y'(e')$ and is a homotopy equivalence of these subcomplexes.

It is easy to show by induction on the number of cells that cell equivalence is an equivalence relation. In fact, given a cell equivalence $h: Y \longrightarrow Y'$ it is not difficult to construct a cell equivalence $g: Y' \to Y$ which is a homotopy inverse of h. It is also easy to see that a cell equivalence preserves the partial order $\leqslant$.

(11.7) THEOREM [F5]. *If f_t is a gradient-like flow on M then there exists a CW complex Y, unique up to cell equivalence, and a homotopy equivalence $h: M \to Y$ such that, for each rest point p of index k, $h(W^u(p))$ is contained in the base $Y(e)$ of a single k-cell e.*

In this way h establishes a one-to-one correspondence between rest points of f_t of index k and k-cells of Y. Moreover, the partial order $\leqslant$ on rest points defined by $q \leqslant p$ if and only if $W^s(q) \cap W^u(p) \neq \varnothing$ corresponds to the partial order $\leqslant$ on cells of Y.

PROOF. The result is valid when M has boundary provided the vector field points inward on the boundary. The map h and the complex Y are defined inductively by skeleta. Choose a self-indexing Morse function g as in (2.8) and let $M_k = g^{-1}([0, k + \frac{1}{2}])$. Suppose $h: M_{k-1} \to Y_{k-1}$ has been defined (the definition when $k - 1 = 0$ being obvious). There is a deformation retraction $r_k: M_k \to M_{k-1} \cup \{D_i\}$ where for each critical point p_i of index k

$$D_i = W^u(p_i) \cap \overline{(M - M_{k-1})}$$

is a k-disk (see (2.9)).

Given $\epsilon > 0$, r_k can be chosen so that if $x \in M_k - M_{k-1}$ is not within ϵ of a stable manifold of a rest point of index k then $r_k(x)$ is the unique point in ∂M_{k-1} on the same orbit as x.

Since for any two rest points p and q, $\overline{W^u(p)} \cap \overline{W^s(q)} \neq \varnothing$ if and only if $W^u(p) \cap W^s(q) \neq \varnothing$ (see the remark after (8.2) of [S1]), it is not difficult to show that there is a $\delta > 0$ such that, for any rest points p and q, $W^u(p) \cap W^s(q) \neq \varnothing$ if and only if there is a point x whose orbit is within δ of both p and q. Now form Y_k by adding k-cells to Y_{k-1}, one for each critical point p_i of index k, with attaching maps $h|\partial D_i$ (as above). Clearly h extends to $h: M_{k-1} \cup \{D_i\} \to Y_k$ and hence to M_k by composing with the retraction r_k. That the new g is a homotopy equivalence is clear.

By choosing the ϵ above small with respect to δ we can assure that, for each $x \in M_k - M_{k-1}$, $r_k(x)$ is on the same orbit as x unless the orbit of x passes within δ of a rest point p of index k in which case the orbit of $r_k(x)$ also passes within δ of p. From this it follows that if p and q are rest points with corresponding cells e and e' in Y, then $e' \leqslant e$ implies $q \leqslant p$. The implication in the other direction is immediate from the definitions.

The uniqueness of Y follows from a consideration of the choices made. First the self-indexing Morse function is unique up to an orbit and level preserving diffeomorphism of the domain M and diffeomorphism of the range R (see (3.4) and (3.13) of [M1]). The construction of the retractions as in (2.9) depends only on the flow and the choice of local coordinates for the standard form of X near rest points. It is easy to see that changes in these can alter the attaching map of a cell e to $Y(e)$ − int e only by a homotopy.   Q.E.D.

We can now give the main result of this chapter showing the correspondence of a connecting manifold with the homotopy class of a relative attaching map via the Pontryagin-Thom construction. Recall that the *relative attaching map* of a k-cell e_k to a j-cell e_j is defined to be the composition

$$S^{k-1} = \partial e_k \xrightarrow{\gamma} Y_j \longrightarrow Y_j/Y_j - \text{int } e_j = S^j$$

where γ is the map attaching e_k to the j-skeleton Y_j and the second map collapses $Y_j - \text{int } e_j$ to a point.

(11.8) THEOREM [F5]. *Suppose p and q are successive rest points of a gradient-like Morse-Smale flow and their connecting manifold is the framed manifold $N \subset S^u(p) \subset W^u(p)$. If Y is the CW complex associated to the flow and α is the relative attaching map of the cell in Y corresponding to p to the cell corresponding to q, the homotopy class corresponding to N under the Pontryagin-Thom construction is the same as the homotopy class of α.*

PROOF. There are coordinates $x_1, \ldots, x_k, y_1, \ldots, y_{n-k}$ on a neighborhood U of q such that the vector field of the flow

$$X = \sum x_i \frac{\partial}{\partial x_i} - \sum y_j \frac{\partial}{\partial y_j}.$$

There is a self-indexing Morse function g with respect to which f_t is gradient-like (see (2.8)), and such that $g(\vec{x}, \vec{y}) = k - \Sigma x_i^2 + \Sigma y_j^2$ on U. We will shrink U if necessary so that

$$U = \{(\vec{x}, \vec{y}) | - \epsilon^2 \leqslant -|\vec{x}|^2 + |\vec{y}|^2 \leqslant \epsilon^2 \text{ and } |\vec{x}||\vec{y}| \leqslant \delta\}$$

for some small $\epsilon, \delta > 0$.

Let $W_0 = g^{-1}([0, k - \epsilon])$ and $W_1 = g^{-1}([0, k + \epsilon])$. Then by (2.9) there is a retraction $r: W_1 \longrightarrow W_0 \cup (\bigcup_i W^u(q_i))$ where $\{q_i\}$ are the critical points of index k and $q = q_1$. Moreover by the construction of r (see (2.7)) if ϵ, δ are sufficiently small then r restricted to U is given by $r(\vec{x}, \vec{y}) = (\vec{x}, \vec{0})$. Let $S^u = W^u(p) \cap g^{-1}(k + \epsilon)$ and let $S^s = W^s(q) \cap g^{-1}(k + \epsilon) \subset U$. The spheres S^u and S^s have a transversal intersection N.

Now by the construction of the associated CW complex Y the attaching map α is homotopic to the composition β

$$S^u \xrightarrow{r} W_0 \cup \left(\bigcup W^u(q_i)\right) \xrightarrow{j} S^k$$

where j collapses $W_0 \cup (\bigcup_{i \neq 1} W^u(q_i))$ to a point and S^k is the one point compactification of the disk $W^u(q) - W_0$. Because S^u is transverse to S^s, it is clear that q is a regular value of β, and that $\beta^{-1}(q) = N$. Also because $r(\vec{x}, \vec{y}) = (\vec{x}, \vec{0})$ on U it is clear that the framing of N given by the normal bundle of $W^s(q)$ is the same as the framing obtained from β. Thus the framed connecting manifold N corresponds to the homotopy class of α via the Pontryagin-Thom construction. Q.E.D.

If q is a successor of p for a flow then of course p is a successor of q for the backwards flow. The connecting manifold for this flow is the same manifold N but with a different framing which comes from the restriction of the normal bundle of $W^u(p)$ (with respect to the original flow). In general these framings of N represent different homotopy classes; however if the manifold M is stably parallelizable then the two homotopy classes are stably the same. The following result should be viewed as a kind of Poincaré duality theorem for stable homotopy. Like the standard Poincaré duality it says that stable and unstable manifolds of a gradient-like flow on a manifold M which is appropriately oriented for the theory give rise to dual cell decompositions.

(11.9) PROPOSITION [F5]. *If p and q are successive rest points, neither of which is a source or sink, of a Morse-Smale flow on a compact stably parallelizable manifold M, then their framed connecting manifold and their connecting manifold framed using the inverse flow determine the same stable homotopy class up to sign.*

PROOF. If we construct the connecting manifold $N = S^s(q) \cap S^u(p)$ we have the following commutative diagram of embeddings with framings

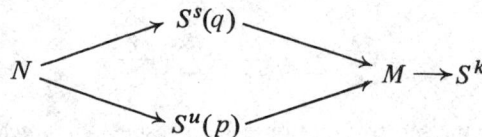

where k is very large. We thus obtain a framing of N in S^k whose homotopy class we wish to show is an iterated (topological) suspension of the homotopy class corresponding to N framed in $S^s(q)$ (or $S^u(p)$). Since neither p nor q is a source or sink the dimension of N is less than $r = \dim S^s(q)$. Thus any two framings of $S^s(q)$ in S^k will result in the same framing of N in S^k (since the difference of the two framings is represented by an element of $\Pi_r(SO(k-r))$ which will restrict trivially to N since the inclusion of N in $S^s(q)$ is null-homotopic). If k is sufficiently large then any two embeddings of $S^s(q)$ are isotopic so it is possible to choose the framing of $S^s(q)$ in S^k in such a way that N framed in S^k will represent the suspension of the element represented by N framed in $S^s(q)$. Since everything said above holds equally well with $S^s(q)$ replaced by $S^u(p)$ the result follows.   Q.E.D.

We conclude this discussion by noting that any stably parallelizable manifold can occur as a connecting manifold, in fact with any stable framing.

(11.10) PROPOSITION [F5]. *Given N a framed submanifold of S^r, there exists an $n > r$, a manifold with boundary M and a gradient-like Morse-Smale vector field X on M pointing inward on the boundary with only three rest points: a sink and p and q such that q is a successor of p and the connecting manifold between p and q is the framed manifold $N \subset S^n$ where $S^r \subset S^n$ is the standard inclusion with standard framing.*

A proof can be found in [F5].

Appendix A. Subshifts as basic sets

In this appendix we give proofs of (3.9), the theorem of Williams characterizing topological conjugacy for subshifts of finite type, and a theorem of Bowen which shows that zero dimensional basic sets are subshifts of finite type (a special case of (3.14)).

(A.1) THEOREM [Wm1]. *Suppose A and B are nonnegative square integer matrices and $\sigma(A)$, $\sigma(B)$ are the corresponding subshifts of finite type. Then $\sigma(A)$ is topologically conjugate to $\sigma(B)$ if and only if A is strong shift equivalent to B.*

PROOF. If $A_1 = RS$, $A_2 = SR$ for nonnegative integer matrices S, R (not necessarily square) we say there is an *elementary shift equivalence* from A_1 to A_2, or $A_1 \sim A_2$. The matrix A is *strong shift equivalent* to B if there are $A = A_1, \ldots, A_n = B$ such that $A_i \sim A_{i+1}$ for $1 \leqslant i < n$. To prove strong shift equivalence implies topological conjugacy of the corresponding shifts it clearly suffices to give the proof when $A \sim B$.

Thus we assume $A = RS$, $B = SR$ and consider the matrix

$$C = \begin{pmatrix} 0 & R \\ S & 0 \end{pmatrix} \quad \text{so that} \quad C^2 = \begin{pmatrix} A & 0 \\ 0 & B \end{pmatrix}.$$

Let $\sigma = \sigma(C) \colon \Sigma_C \to \Sigma_C$. Since by (3.3), $\sigma(C)^2$ is topologically conjugate to $\sigma(C^2)$, Σ_C is the disjoint union of sets X and Y with $\sigma^2 | X$ topologically conjugate to $\sigma(A)$ and $\sigma^2 | Y$ topologically conjugate to $\sigma(B)$. From the form of C it is easy to see that σ maps X to Y and Y to X. Thus $\sigma | X \colon X \to Y$ is a topological conjugacy between $\sigma^2 | X$ and $\sigma^2 | Y$, so $\sigma(A)$ and $\sigma(B)$ are topologically conjugate.

We prove the converse in a sequence of lemmas.

(A.2) LEMMA. *If A is an $n \times n$ nonnegative integer matrix then A is strong shift equivalent to a matrix A' whose entries are zeroes and ones.*

PROOF. Let G be the edge graph corresponding to A and suppose its vertices are $v(1), \ldots, v(n)$. At the center of each edge of G place a new vertex and label these vertices $u(1), \ldots, u(m)$, where $m = \Sigma_{ij} A_{ij}$ is the number of edges. Define matrices R, S by

$$R_{ij} = \begin{cases} 1 & \text{if } u(j) \text{ is on the edge emanating from } v(i), \\ 0 & \text{otherwise} \end{cases}$$

and

$$S_{kl} = \begin{cases} 1 & \text{if } u(k) \text{ is on the edge ending in } v(l), \\ 0 & \text{otherwise.} \end{cases}$$

The matrices R and S have size $n \times m$ and $m \times n$ respectively. Now $(SR)_{kj} = \Sigma_i S_{ki} R_{ij} = 0$ or 1, since there is at most one i with $v(i)$ the end of the edge containing $u(k)$ and the beginning of the edge containing $u(j)$. Thus SR is a matrix of zeroes and ones. We also have $(RS)_{il} = \Sigma_j R_{ij} S_{jl} =$ number of edges going from $v(i)$ to $v(l)$ since $R_{ij} S_{jl} = 1$ if $u(j)$ is on an edge going from $v(i)$ to $v(l)$ and 0 otherwise. Thus $RS = A$. Q.E.D.

If $\sigma(A)$ and $\sigma(B)$ are topologically conjugate we want to show A and B are strong shift equivalent. Lemma (A.2) and the fact that strong shift equivalent matrices correspond to topologically conjugate shifts allows us to restrict our attention to matrices A and B of zeroes and ones and their corresponding vertex shifts (see (3.3)) $\sigma(A)$ and $\sigma(B)$.

Suppose $\sigma: \Sigma \longrightarrow \Sigma$ is the vertex shift corresponding to some matrix of zeroes and ones. There is a metric ρ defined on Σ by $\rho(\mathbf{c}, \mathbf{d}) = \Sigma_{n=-\infty}^{\infty} 2^{-|n|} \delta(c_n, d_n)$, where $\delta(c_n, d_n) = 1$ if $c_n = d_n$ and 0 otherwise. We can define stable and unstable "manifolds" for σ and also the concepts of rectangle and Markov partition. Thus $W^s(\mathbf{c})$ is defined to be $\{\mathbf{d} | d_i = c_i \text{ when } i \leq n, \text{ for some } n\} = \{\mathbf{d} | \rho(\sigma^m(\mathbf{c}), \sigma^m(\mathbf{d})) \to 0 \text{ as } m \to \infty\}$ and $W^u(\mathbf{c})$ is defined similarly. We will also use local stable and unstable manifolds defined by $W_0^s(\mathbf{c}) = \{\mathbf{d} | c_i = d_i, i \leq 0\}$ and $W_0^u(\mathbf{c}) = \{\mathbf{d} | c_i = d_i, i \geq 0\}$.

If $\mathbf{c}$ and $\mathbf{d}$ are sufficiently close, in fact if $c_0 = d_0$ then $W_0^s(\mathbf{c}) \cap W_0^u(\mathbf{d})$ consists of the single point $\mathbf{z}$ with $z_i = c_i$, $i \leq 0$, and $z_i = d_i$, $i \geq 0$. Hence whenever $c_0 = d_0$ we can define $[\mathbf{c}, \mathbf{d}]$ to be the point $W_0^s(\mathbf{c}) \cap W_0^u(\mathbf{d})$. If $R \subset \Sigma$ we say R is a *rectangle* if it is open and closed, and if $\mathbf{c}, \mathbf{d} \in R$ implies $c_0 = d_0$ and $[\mathbf{c}, \mathbf{d}] \in R$. If R is a rectangle and $\mathbf{c} \in R$ define $W^s(\mathbf{c}, R)$ to be $W_0^s(\mathbf{c}) \cap R$ and define $W^u(\mathbf{c}, R)$ to be $W_0^u(\mathbf{c}) \cap R$.

A *Markov partition* for $\sigma: \Sigma \longrightarrow \Sigma$ is a finite covering $\{R_1, \dots, R_m\}$ of Σ by disjoint rectangles such that if $\mathbf{c} \in R_i$, $\sigma(\mathbf{c}) \in R_j$ then

$$\sigma(W^u(\mathbf{c}, R_i)) \supset W^u(\sigma(\mathbf{c}), R_j) \quad \text{and} \quad \sigma(W^s(\mathbf{c}, R_i)) \subset W^s(\sigma(\mathbf{c}), R_j).$$

If $\sigma: \Sigma \longrightarrow \Sigma$ is the vertex shift associated to the $n \times n$ matrix A with vertices labelled 1 to n, and $U(A) = \{R_i\}$, where $R_i = \{\mathbf{c} | c_0 = i\}$, then $U(A)$ is a Markov partition for σ.

Conversely if $U = \{R_1, \dots, R_n\}$ is a Markov partition for $\sigma: \Sigma \longrightarrow \Sigma$ then we can associate to it a matrix $M = M(U)$ of zeroes and ones defined by $M_{ij} = 1$ if $R_i \cap \sigma(R_j) \neq \varnothing$ and 0 otherwise. Clearly $M(U(A)) = A$.

Exercise. Show if $M = M(U)$ where U is a Markov partition for $\sigma: \Sigma \longrightarrow \Sigma$, then σ is topologically conjugate to $\sigma(M): \Sigma_M \longrightarrow \Sigma_M$.

From the definitions it is easy to see that if U and V are Markov partitions then so are $U \cap V$, $U \cap \sigma(V)$, and $\sigma^{-1}(U) \cap V$, where $U \cap V$ denotes the partition $\{R_i \cap R_j'\}$, $R_i \in U$, $R_j' \in V$. We will say U *refines* V (denoted $U > V$) if each element of U is contained in an element of V. Given a Markov partition U and integers $m, n \geq 0$ let $U(-m, n)$ denote the Markov partition $\sigma^{-m}(U) \cap \cdots \cap \sigma^n(U)$.

(A.3) LEMMA. *Suppose U and V are Markov partitions for $\sigma: \Sigma \longrightarrow \Sigma$ with associated matrices $A = M(U)$ and $B = M(V)$. If $U > V$ and either $V(0, 1) > U$ or $V(-1, 0) > U$ then there are matrices of zeroes and ones, R and S with $A = SR$ and $B = RS$.*

PROOF. Suppose $V(0, 1) > U$, the other case being similar. If $U = \{u_1, \ldots, u_n\}$ and $V = \{v_1, \ldots, v_m\}$ define matrices R and S of size $n \times m$ and $m \times n$ respectively by

$$R_{ij} = \begin{cases} 1 & \text{if } u_j \subset v_i, \\ 0 & \text{otherwise} \end{cases}$$

and

$$S_{kl} = \begin{cases} 1 & \text{if } v_r \cap \sigma(v_l) \subset u_k \text{ where } v_r \text{ is the element of } V \text{ containing } u_k, \\ 0 & \text{otherwise.} \end{cases}$$

Now $B_{il} = 1$ if and only if $v_i \cap \sigma(v_l) \neq \varnothing$, and $(RS)_{il} = \Sigma_j R_{ij} S_{jl} = 1$ if $v_i \cap \sigma(v_l) \neq \varnothing$ since there is then a unique u_j satisfying $v_i \cap \sigma(v_l) \subset u_j \subset v_i$, and $R_{ij} S_{jl} = 1$ if and only if such a u_j exists. If $v_i \cap \sigma(v_l) = \varnothing$ one of R_{ij} and S_{jl} must vanish so $(RS)_{il} = 0$. Hence we have shown $B = RS$.

Similarly $A_{kj} = 1$ if and only if $u_k \cap \sigma(u_j) \neq \varnothing$ and $(SR)_{kj} = \Sigma_i S_{kl} R_{lj} = 1$ if and only if $u_k \cap \sigma(u_j) \neq \varnothing$ since then if $v_i \supset u_j$ and $v_r \supset u_k$ we have $R_{ij} = 1$ ($R_{lj} = 0$, if $l \neq i$) and $v_r \cap \sigma(v_i)$ is nonempty and contained in u_k so $S_{ki} = 1$. It is also clear that if $u_k \cap \sigma(u_j) = \varnothing$, $S_{kl} R_{lj} = 0$ for all l, so $A = SR$.   Q.E.D.

(A.4) LEMMA. *If U and V are Markov partitions for $\sigma: \Sigma \longrightarrow \Sigma$ and $A = M(U)$, $B = M(V)$ are the associated matrices of zeroes and ones, then A and B are strong shift equivalent.*

PROOF. If U and U' are two Markov partitions with $M(U)$ strong shift equivalent to $M(U')$ we will write $U \sim U'$. Using induction it is an immediate corollary of (A.3) that $U \sim U(-m, n) = \sigma^{-m}(U) \cap \cdots \cap \sigma^n(U)$ for all $m, n \geqslant 0$. It is not difficult to show that for N sufficiently large $U(-N, N)$ refines V (use the fact that balls defined in the metric ρ are open and closed and any two of them of the same radius are either equal or disjoint). Thus replacing U by $U(-N, N)$ (which we call U from now on) we can assume $U > V$.

For some $m, n \geqslant 0$ we will also have $V(-m, n) > U$. We prove by induction that $U \sim V(-m, n)$ which implies $U \sim V$ since $V \sim V(-m, n)$. The induction hypothesis is that whenever U, V are Markov partitions such that $U > V$, $V(-m, n) > U$ and $m + n \leqslant k$ then $U \sim V(-m, n)$. Lemma (A.3) says this hypothesis is valid when $k = 1$. We now assume it true for k and prove it for $k + 1$.

Thus suppose we are given U, V such that $U > V$, $V(-m, n) > U$ and $m + n = k + 1$. Let $W = U \cap \sigma(V)$ so $W > U$ and $U(0, 1) > W$ since $\sigma(U) > \sigma(V)$. Thus $U \sim W$ by Lemma (A.3).

Either $m \geqslant 1$ or $n \geqslant 1$; we assume $n \geqslant 1$, the other case being similar. Then $V(-m, n)$ $> \sigma(V)$ and $V(-m, n) > U$ so $V(-m, n) > W$. Also $W(-m, n-1) = U(-m, m-1) \cap$ $\sigma(V(-m, n-1)) = U(-m, n-1) \cap V(-m+1, n)$ so $W(-m, n-1) > \sigma^{-m}(U) > \sigma^{-m}(V)$ and $W(-m, n-1) > V(-m+1, n)$. Thus $W(-m, n-1) > V(-m, n)$. Since $V(-m, n) > W$ and $W(-m, n-1) > V(-m, n)$ and $m + n - 1 = k$, the induction hypothesis implies $W \sim$ $V(-m, n)$ so $U \sim W \sim V(-m, n)$ which proves the induction hypothesis for $k + 1$. Since $U \sim V(-m, n) \sim V$ this completes the lemma.

We can now complete the proof of the theorem. Suppose A and B are matrices of zeroes and ones and $h: \Sigma_B \longrightarrow \Sigma_A$ is a topological conjugacy between $\sigma(B)$ and $\sigma(A)$. Let U be a Markov partition for $\sigma(A)$ with $A = M(U)$ (e.g., U the standard partition $\{R_i = \{c \in \Sigma_B | c_0 = i\})$, and V a Markov partition for $\sigma(B)$ with $M(V) = B$. For N sufficiently large $h(V(-N, N))$ will be a Markov partition for $\sigma(A)$. Thus $V \sim V(-N, N)$ and $U \sim$ $h(V(-N, N))$, but the matrices associated to $V(-N, N)$ and $h(V(-N, N))$ are clearly the same. Thus A is strong shift equivalent to B.

By (A.2) it is sufficient to consider the case when A and B are matrices of zeroes and ones. Q.E.D.

The general proof of Bowen's Theorem (3.14) is somewhat involved and is well done in [B2]. We give a proof from [B5] of the much easier special case when Λ is zero dimensional.

(A.5) THEOREM. *If Λ is a zero dimensional basic set of a diffeomorphism $f: M \longrightarrow M$ with hyperbolic chain recurrent set then $f|\Lambda$ is topologically conjugate to a subshift of finite type.*

PROOF. First observe that $f|\Lambda$ is *expansive* i.e., there is an $\epsilon > 0$ such that, if $x, y \in \Lambda$ and $d(f^n(x), f^n(y)) < \epsilon$ for all n then $x = y$. This follows from the fact that $d(f^n(x), f^n(y)) < \epsilon$ implies $y \in W_\epsilon^s(x)$ and $y \in W_\epsilon^u(x)$, but our remarks defining rectangles indicated that if ϵ is sufficiently small $W_\epsilon^s(x) \cap W_\epsilon^u(x)$ consists of a single point $[x, x] = x$. Since Λ is zero dimensional it is possible to cover it by a finite collection C of disjoint sets each of which is open and closed in Λ and of diameter $< \epsilon$. Give C the discrete topology and define $h: \Lambda \longrightarrow \Pi_{-\infty}^\infty C$ by $h(x) = c = (\ldots, c_{-1}, c_0, c_1, \ldots)$ where $f^{-i}(x) \in c_i$ for all i. Clearly h is continuous and $h \circ f = \sigma \circ h$ where $\sigma: \Pi_{-\infty}^\infty C \longrightarrow \Pi_{-\infty}^\infty C$ is the shift to the right. The expansiveness of f implies that if $x \neq y$ then $d(f^n(x), f^n(y)) > \epsilon$ for some n. Hence $h(x)$ and $h(y)$ must differ in place $-n$, so h is one-to-one. Thus if $X = h(\Lambda)$, $\sigma:$ $X \longrightarrow X$ is a "subshift" and we must show it is of finite type.

We can define stable and unstable "manifolds" in X by $W^s(c) = \{d|$ for some n, $c_i = d_i$ if $i \geqslant n\}$, $W^u(c) = \{d|$ for some n, $c_i = d_i$ if $i \leqslant n\}$ and local stable and unstable manifolds by $W_0^s(c) = \{d|d_i = c_i$ for $i \geqslant 0\}$, $W_0^u(c) = \{d|d_i = c_i$, for $i \leqslant 0\}$. The continuity of h together with the fact that for $x, y \in \Lambda$ the function $(x, y) \longrightarrow [x, y] = W_\epsilon^s(x) \cap W_\epsilon^u(y)$ described above is continuous, implies that we can define $[c, d] \in X$ if c and d are sufficiently close. Specifically, there exists $N > 0$ such that if $c, d \in X$ and $c_i = d_i$, $-N \leqslant i \leqslant N$, then we can define $[c, d]$ to be z where $z_i = c_i$, $i \geqslant 0$, $z_i = d_i$, $i \leqslant 0$, and we will have

$[c, d] \in X$. We know $[c, d] \in X$ because if $c = h(x)$, $d = h(y)$ and x and y are sufficiently close (i.e., N is large enough) then $h([x, y]) \in W_0^s(c) \cap W_0^u(d)$, so $h([x, y]) = [c, d]$.

To prove that $\sigma: X \longrightarrow X$ is conjugate to a subshift of finite type it is enough to show that if $n = 2N + 2$ there is a subset S of $C^n = \{(c_1, \ldots, c_n) | c_i \in C\}$ such that $c \in \Pi_{-\infty}^{\infty} C$ is in X if and only if $(c_{k+1}, \ldots, c_{k+n}) \in S$ for all k. (This is because σ is then conjugate to the shift determined by S and $\longrightarrow$ when we define $(c_1, \ldots, c_n) \longrightarrow (c_1', \ldots, c_n')$ if and only if $c_i = c_{i-1}'$, $2 \leqslant i \leqslant n$.) The set S is defined to be the set of all n-blocks which occur in elements of X, i.e., since X is invariant under σ,

$$S = \{(c_1, \ldots, c_n) \in C^n | \text{ there is } d \in X \text{ with } d_i = c_i, 1 \leqslant i \leqslant n\}.$$

Clearly if $c \in X$ every n-block in c is in S. We prove the converse by showing if $c \in \Pi_{-\infty}^{\infty} C$ and every n-block of c is in S then for each $m > 0$ there is an element $c^m \in X$ such that $c_i^m = c_i$ for $-m \leqslant i \leqslant m$. Thus since X is closed and $\lim_{m \to \infty} c^m = c$, we will have $c \in X$. We produce c^m by induction on m; if $2m \leqslant n$, c^m exists by the definition of S. Our induction hypothesis is that for every c a corresponding c^m exists. We then define $c^{m+1} = [\sigma^{-1}((\sigma(c))^m), \sigma((\sigma^{-1}(c))^m)]$ which is in X, since $(\sigma^{-1}(c))^m$ and $(\sigma(c))^m$ are, and which agrees with c in places $-m - 1$ to $m + 1$. Thus $c = \lim c^m$ is in X and hence $\sigma: X \longrightarrow X$ is topologically conjugate to a subshift of finite type.   Q.E.D.

Appendix B. Constructing fitted diffeomorphisms

In this appendix we give the proofs of Theorems (4.6) and (4.7) and of Proposition (4.8).

(B.1) LEMMA. *Let g_t be a flow on M gradient-like with respect to a self-indexing Morse function ϕ and let $\{H(k)\}$ be the associated handle sets. Then a diffeomorphism f_0: $M \to M$ can be isotoped to f satisfying*

(i) *For each k, f is hyperbolic on handle set $H(k)$.*

(ii) *For each k, $f(M_k) \subset \text{int}(M_{k-1} \cup H(k))$, where $M_k = \phi^{-1}((-\infty, k + 1 - \epsilon])$.*

Choosing ϕ appropriately the isotopy can be made C^0 small.

PROOF. We first alter f_0 by a small isotopy to f_1 so that $f_1(W^u(p_i^k))$ is transverse to $W^s(p_j^l)$ (denoted $f_1(W^u(p_i^k)) \pitchfork W^s(p_j^l)$) for all p_i^k, p_j^l, the centers of $h_i(k) \in H(k)$ and $h_j(l) \in H(l)$ respectively. Here and henceforth $W^u(p_i^k)$ denotes the unstable manifold with respect to the flow g_t, not f_0 or f_1. We will continue to denote k-disks in $h_i(k) = D_i^k \times D_i^{n-k}$ by $W_i^u(x)$, $x \in h_i(k)$, as before.

We now alter f_1 to achieve (ii). Let $f_2 = g_T \circ f_1 \circ g_T$ where $T > 0$. We will show by induction on k that if T is sufficiently large f_2 satisfies (ii). For $k = 0$ this is clear since the transversality condition on f_1 implies $f_1(\{p_i^0\}) \subset \bigcup_i W^s(p_i^0)$ and $g_T(M_0)$ is a small neighborhood of $\{p_i^0\}$ if T is large. We now assume the result true for $k \leqslant K - 1$ and prove it for K. For t sufficiently large $g_t(M_K) \subset \text{int}(M_{K-1} \cup H(K))$; hence we need only show $g_t \circ f_1 \circ g_t(H(K)) \subset \text{int}(M_{k-1} \cup H(K))$ for large t. But the transversality condition on f_1 implies $g_t \circ f_1(W_i^u(K)) \subset \text{int}(M_{K-1} \cup H(K))$ for large t, since $f_1(W_i^u(K)) \cap W^s(p_j^l)$ is empty if $l > K$. It follows that if U is a small neighborhood in $H(K)$ of $\{W_i^u(K)\}$ then $g_t \circ f_1(U) \subset \text{int}(M_{K-1} \cup H(K))$ for large t. Since for large t, $g_t(H(K)) \subset U \cup M_{K-1}$ it follows that $g_t \circ f_1 \circ g_t(H(K)) \subset \text{int}(M_{K-1} \cup H(K))$. By the induction hypothesis $g_t \circ f_1 \circ g_t(M_{K-1}) \subset \text{int } M_{K-1}$ so we have (ii) for K.

We now alter f_2 to obtain (i). We already have $f_2(W_j^u(k)) \pitchfork W_i^s(k)$ for all i, j, k; hence if $U_j(k)$ is a sufficiently small neighborhood of $W_j^u(k)$ in $h_j(k)$ we will have $f_2(W_j^u(x)) \pitchfork W_i^s(k)$ for all $x \in U_j(k)$. Since for large t, $g_t(h_j(k)) \subset U_j(k) \cup M_{k-1}$, if we let $f_3 = f_2 \circ g_t$ we will have $f_3(W_j^u(x)) \pitchfork W_i^s(k)$ for all $x \in h_j(k)$ and all i. Let $D = f_3(h_j(k)) \cap W_i^s(k)$ so D is a finite set of $(n - k)$-disks in $W_i^s(k)$. It is not difficult to write down an explicit isotopy supported on a small neighborhood of D which for each $x \in D$ flattens a neighborhood of x in the k-disk $f_3(W_j^u(f_3^{-1}(x))$ along the $(n - k)$-disks $\{W_i^s(y)\}$ onto a neighborhood $V_{ij}(x)$ of

x in the k-disk $W_i^u(x)$ (see Figure (B.1)). Call the new diffeomorphism obtained by following f_3 with this isotopy f_4.

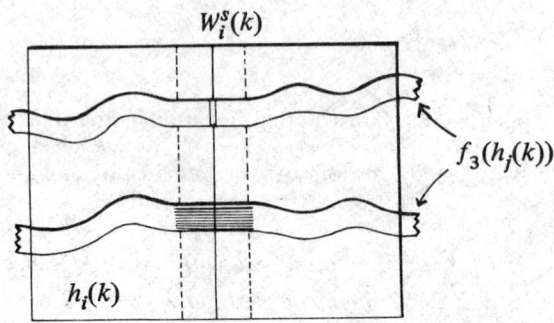

FIGURE (B.1)

The result is that whenever $y \in V_{ij}(x)$, $f_4(W_j^u(f_4^{-1}(y))) \subset V_{ij}(x)$. We now choose t sufficiently large that, for each x, g_t spreads $V_{ij}(x)$ all the way across $h_i(k)$; specifically we need that $g_t(V_{ij}(x)) \supset W_i^u(g_t(x))$ for all $x \in W_i^s(k)$.

Then if $f_5 = g_t \circ f_4$ we will have $f_5(W_j^u(f^{-1}(x)) \supset W_i^u(x)$. The argument is repeated for each pair of k-handles $h_i(k)$, $h_j(k)$ and for each handle set $H(k)$, which gives part of the property of being hyperbolic on $H(k)$.

We digress briefly before completing the proof.

(B.2) REMARK. For convenience we defined handles in Chapter 2 which are twice as big in the W^u direction as the W^s direction, i.e., $h_i(k) = \{(\vec{x}, \vec{y}) \mid |\vec{x}|^2 \leq 2\epsilon, |\vec{y}|^2 \leq \epsilon\}$, in local coordinates centered at a critical point p_i^k of index k. It is not at all necessary that the handles have this shape, however, and sometimes it is convenient to alter the shape. Thus if $0 < a$, $b < 1$, we might define $\tilde{h}_i(k) = \{(\vec{x}, \vec{y}) \mid |\vec{x}|^2 \leq 2a\epsilon, |\vec{y}|^2 \leq b\epsilon\} \subset h_i(k)$ and if f is hyperbolic on the handle set $H(k) = \{h_i(k)\}$, then for t sufficiently large $\tilde{f} = g_t \circ f \circ g_t$ will be hyperbolic on $\tilde{H}(k) = \{\tilde{h}_i(k)\}$. Likewise if some of the defining properties of fitted diffeomorphisms hold for f and the handle sets $H(k)$, the same properties will hold for $\tilde{f}$ and $\tilde{H}(k)$ if t is sufficiently large. This allows us to use handles of a different shape provided we replace f with $\tilde{f}$.

We return now to the proof of (B.1) where we apply precisely the same argument to f^{-1} using g_{-t} to obtain $f^{-1}(W_j^s(f(x))) \supset W_i^s(x)$ whenever $x \in f^{-1}(h_j(k)) \cap h_i(k)$. (We must actually redefine $h_i(k)$ to be $\{(\vec{x}, \vec{y}) \mid |\vec{x}|^2 \leq \frac{1}{2}\epsilon, |\vec{y}|^2 \leq \epsilon\}$ and alter f as just described in (B.2).)

Notice that the flattening is done along $\{W_i^u(y)\}$ so we do not destroy the inclusion $f(W_j^u(x)) \supset W_i^u(f(x))$. If f does not satisfy the expanding and contracting condition for hyperbolic handles, then replacing it one more time by $g_t \circ f \circ g_t$ will achieve this. To make all of these isotopies C^0 small choose a triangulation of M with very small simplices and construct g_t with a rest point of index k at the barycenter of each k-simplex. Consequently g_t

will be C^0 close to the identity for all $t \geqslant 0$. The other deformations were chosen to be C^0 small. Q.E.D.

(B.3) LEMMA [SS]. *If $\phi: M \longrightarrow R$ is a self-indexing Morse function, there is a flow g_t gradient-like with respect to ϕ and a collection of handle sets $\{H(k)\}$ such that the time T map of the flow g_T is fitted with respect to $\{H(k)\}$ for all T sufficiently large.*

PROOF. We first choose a gradient-like flow g_t^0 which satisfies the transversality condition (cf. (2.8)), and handle sets $H(k) = \{h_i(k)\}$ given by $h_i(k) = \{(\vec{x}, \vec{y}) \mid |\vec{x}|^2 \leqslant 2\epsilon, |\vec{y}|^2 \leqslant \epsilon\}$ in local coordinates $(\vec{x}, \vec{y})$ centered at the critical point p_i^k of index k. Let $M_k = \phi^{-1}((-\infty, k + 1 - \epsilon])$ as before.

Any gradient-like flow automatically satisfies condition (a) (that it be hyperbolic on the handle sets $H(k)$) of the definition of fitted. Also if T is sufficiently large $g_T^0(M_k) \subset M_{k-1} \cup H(k)$. Thus we want to alter g_t^0 so that for all $t \geqslant 0$ and $x \in h_i(k)$, $y = g_t^0(x) \in h_j(l)$ we have $g_t^0(W_i^u(x)) \supset W_j^u(y)$.

Suppose inductively we have achieved this whenever $k \leqslant K - 1$ (it is trivial when $k = 0$). To achieve it for K we will make use of a new self-indexing Morse function ϕ_0 which is obtained by altering ϕ slightly so that for each $h_i(K) = D_i^K \times D_i^{n-K}$, the set $\partial^u(h_i(K)) = (\partial D_i^K) \times D_i^{n-K}$ is contained in $\phi_0^{-1}(K - \epsilon)$ and for each $h_j(l)$, $l < K$, the set $\partial^s(h_j(l)) = D_j^l \times \partial D_j^{n-l}$ is contained in $\phi_0^{-1}(l + \epsilon)$. This is not difficult (cf. (B.2) and the remarks after (2.6)) and can be done so that g_t is gradient-like with respect to ϕ_0. We do not change the flow g_t or the handles $\{H(k)\}$.

We will first alter the flow on $N = \phi_0^{-1}([K - 1 + \epsilon, K - \epsilon])$. By (2.4) if $V = \phi_0^{-1}(K - 1 + \epsilon)$, N is diffeomorphic to $V \times I$ by a diffeomorphism carrying the curve (x, t), $t \in [0, 1]$, to an orbit segment of the flow g_t^0 on N. Suppose now that $\alpha: V \longrightarrow V$ is a diffeomorphism isotopic to the identity (say by a smooth isotopy α_t with $\alpha_t = $ id for $t \in [0, 1/3]$ and $\alpha_t = \alpha$ for $t \in [2/3, 1]$). Then we can define a new flow g_t' whose flow lines on N are the images of the curves $(\alpha_t(x), t)$ on $V \times I$.

Thus if we have parametrized g_t^0 so that $g_1^0(\phi_0^{-1}(K - \epsilon)) = V$ we will have, for each $y \in \phi_0^{-1}(K - \epsilon)$, $g_1'(y) = \alpha(g_1^0(y))$.

We now construct α. If $x \in h_i(K)$ let $S_i^u(x) = g_1^0(\partial W_i^u(x))$ or equivalently $g_1^0(W_i^u(x) \cap \phi_0^{-1}(K - \epsilon))$, so $S_i^u(x)$ is the boundary $(K - 1)$-sphere of $W_i^u(x)$ flowed by g_t^0 down into V. Similarly if $y \in h_j(K - 1)$ let $S_j^s(y) = \partial W_j^s(y) \subset V$, and let $S_j = \partial W_j^s(p_j)$ where p_j is the center of $h_j(K - 1)$. If we now find $\alpha: V \longrightarrow V$ isotopic to the identity with $\alpha(S_i^u(x)) \supset W_j^u(y)$ for all $x \in h_i(K)$, $y \in S_j$, and all i and j, then the flow g_t' constructed from α will satisfy $g_T'(W_i^u(x)) \supset W_j^u(y)$ for sufficiently large T, if $x \in W_i^u(K)$ and $y = g_t^0(x) \in W_j^u(K - 1)$, $t \geqslant 0$.

Let α_1 be a perturbation of the identity satisfying $\alpha_1(W_j^u(y)) = W_j^u(y)$ for $y \in S_j$ and $\alpha_1(S_i^u(p_i^K)) \pitchfork S_j$ for all i, j. It follows that there is a small neighborhood U_i of p_i^k in $h_i(K)$ and a neighborhood Y_j of S_j in $\bigcup_{z \in S_j} W_j^u(z)$ such that whenever $x \in U_i$, $y \in Y_j$, $\alpha_1(S_i^u(x)) \pitchfork S_j(y)$. We now do a flattening isotopy (see Figure (B.2)) isotoping α_1 to α so that $\alpha(S_i^u(x))$ contains $W_j^u(y) \cap Y_j$ if $y \in \alpha(S_i^u(x)) \cap S_j$ and $x \in U_i$. We do this in such a way that $\alpha(S_j^s(z)) = S_j^s(z)$ for all $z \in \partial h_j(K-1)$, i.e., we isotope along the spheres $\{S_j^s(z)\}$.

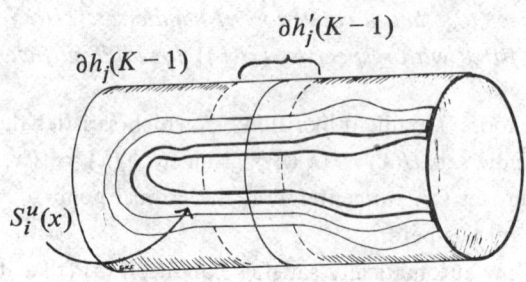

FIGURE (B.2)

We now have the condition we want locally and we make it global by shrinking the handles $h_i(K)$ and $h_j(K-1)$. Specifically if $h_j(K-1) = D_j^{K-1}(\epsilon) \times D_j^{n-K+1}(\epsilon) = \{(\vec{x}, \vec{y})| |\vec{x}|^2 \leqslant \epsilon, |\vec{y}|^2 \leqslant \epsilon\}$ and $h_i(K) = D_i^K(\epsilon) \times D_i^{n-K}(\epsilon)$ choose $\epsilon' < \epsilon$ small enough that if $h_j'(K-1) = D_j^{K-1}(\epsilon') \times D_j^{n-K+1}(\epsilon) = \{(\vec{x}, \vec{y})| |\vec{x}|^2 \leqslant \epsilon', |\vec{y}|^2 \leqslant \epsilon\}$ then for the new handle $W_i^u(y) \subset Y_j$ for all $y \in S_j$. Likewise we choose ϵ' small enough that if $h_i'(K) = D_i^K(\epsilon) \times D_i^{n-K}(\epsilon')$ then, for any $x' \in h_i'(K)$, $W_i^u(x') = W_i^u(x)$ for some $x \in U_i$. With these new handles $\alpha(S_i^u(x))$ will contain $W_j^u(y)$ or be disjoint from it for all $x \in h_i'(K)$ and $y \in h_j'(K-1) \cap V$. This change in the handles does not affect fittedness in levels $K-1$ and lower. After these changes have been made for all i, j, we have a flow g_t' such that if $x \in h_i(K)$ and $y \in h_j(K-1)$ then, for all T, $g_T'(W_i^u(x)) \supset W_j^u(y)$ if they intersect.

We now want to repeat this argument for $x \in h_i(K)$ and $y \in h_k(K-2)$. To do this note that after reparametrizing g_t' we can arrange that if $V_0 = V - (V \cap H(K-1))$ then $g_1'(V_0) \subset V' = \phi_0^{-1}(K-2+\epsilon)$. By our induction hypothesis if $x \in h_i(K)$, $y \in h_k(K-2) \cap (V - g_1'(V_0))$ and $g_t'(x) = y$ for some t, then there are a j and $z \in V \cap h_j(K-1)$ on the orbit from x to y, so we have $g_t'(W_i^u(x)) \supset W_j^u(z)$ for some $t > 0$ and hence $g_s'(W_i^u(x)) \supset W_k^u(y)$ for some $s > t$. Therefore we can repeat the argument constructing $\alpha': V' \to V'$ isotopic to the identity and equal to the identity off of $g_1'(V_0)$. Using α' we alter the flow as before to g_t satisfying $g_T(W_i^u(x)) \supset W_k^u(y)$ if $x \in h_i(K)$ and $y = g_t(x) \in h_k(K-2)$. Continuing in this way we finally achieve our induction hypothesis for K. This gives the condition that $g_t(W_i^u(x)) \supset W_j^u(y)$, if they intersect, for all $t \geqslant 0$ and all x, y in handles. The whole argument must now be repeated for g_{-t} and $\{W_i^s(x)\}$. One should check that the constructions were done so that this does not destroy the fitting of g_t and $\{W_i^u(x)\}$. Q.E.D.

(B.4) PROOF OF (4.6). According to (B.3) we can choose a self-indexing Morse function ϕ with gradient-like flow g_t and handle sets $\{H(k)\}$ such that g_T is fitted with respect to $\{H(k)\}$ for large T. By (B.1) we can isotope f_0 so that it is hyperbolic with respect to the handle sets $\{H(k)\}$ and satisfies $f_0(M_k) \subset \text{int}(M_{k-1} \cup H(k))$ for all k, and from the proof it is clear that the same is true for $f_0(t, t') = g_t \circ f_0 \circ g_{t'}$ whenever $t, t' \geqslant 0$.

It remains to isotope f_0 to f so that in addition whenever $x \in h_i(k)$, $y \in h_j(l)$, $l < k$, and $f^n(x) = y$ we have $f^n(W_i^u(x)) \supset W_j^u(y)$, which we will show is possible by induction on k. Our induction hypothesis is that whenever $k < K$ this condition holds for f_0 and $f_0(t, t')$ $= g_t \circ f_0 \circ g_{t'}$ if $t, t' \geq 0$. This hypothesis is trivially verified if $k = 0$.

Instead of altering f_0 on the K-handles $H(K) = \bigcup_i h_i(K)$ we choose $T > 0$ sufficiently large that $g_T(H(K) \cap M_{K-1}) \subset (H(K-1) \cup M_{K-2})$ and work with the handles $h_i'(K) = g_T(h_i(K))$ whose union we denote $H'(K)$. Note that $g_T \circ f_0 \circ g_{-T}$ is hyperbolic on the handle set $H'(K)$ and hence so is $f_1(t, t') = g_t \circ (g_T \circ f_0 \circ g_{-T}) \circ g_{t'}$, whenever $t, t' \geq 0$. Let $f_1 = f_1(0, T) = f_0 \circ g_T$, then f_1 has all the properties of f_0 but is in addition hyperbolic on the handle set $H'(K)$.

Our plan is to produce a diffeomorphism f_4 with all the properties of f_1 but also such that whenever $x \in h_i'(K)$ and $y = f_4(x) \in h_j(K-1)$ we have $f_4(W_i^u(x)') \supset W_j^u(y)$ (where $W_i^u(x)' = g_T(W_i^u(g_{-T}(x)))$) and such that $g_t \circ f_4 \circ g_{t'}$ has the same properties if $t, t' \geq 0$.

Given such an f_4 we claim that $f = f_4 \circ g_T$ will have the additional property that whenever $x \in h_i(K)$, $y = f^n(x) \in h_j(K-1)$, then $f^n(W_i^u(x)) \supset W_j^u(y)$ which is what we really want (i.e., $H(K)$ instead of $H'(K)$ and f^n, $n \geq 1$, instead of just f). The proof of this claim follows from the fact that T was chosen so that if $z \in H'(K)$ and $f(z) \notin H'(K)$ then $f(z) \in H(K-1) \cup M_{K-2}$. If $x \in h_i(K)$, $y = f^n(x) \in h_j(K-1)$, let x' be the last point on the orbit segment from x to y which is not in $H(K-1)$ (i.e., $x' = f^m(x) \notin H(K-1)$ but $f^{m+1}(x) \in H(K-1)$). Then $x' \in h_l'(K)$ for some l, so

$$f^n(W_i^u(x)) \supset f^{n-m}(W_l^u(x')) \supset f^{n-m-1}(W_{j'}^u(f(x')) \supset W_j^u(y)$$

where $W_{j'}^u(f(x'))$ and $W_j^u(y)$ are the $(K-1)$-disks in $h_{j'}(K-1)$ and $h_j(K-1)$. This completes the proof of the claim.

We proceed with the construction of f_4. Notice that if $x \in h_i'(K) \cap H(l)$, $l < K$, and $y = f_1^n(x) \in h_j(l')$, $l' \leq l$, then we already have $f_1^n(W_i^u(x)') \supset W_j^u(y)$ because of the induction hypothesis and the fact that g_T is fitted with respect to $\{H(k)\}$.

We now perturb f_1 to f_2 so that $f_2(W_i^u(K)') \pitchfork W_j^s(K-1)$ for all i, j. (As before $W_i^u(k)$ denotes $W_i^u(p_i^k)$ where p_i^k is the center of $h_i(k)$.) No change is needed on $H(l)$, $l < K$. By openness of transversality there are neighborhoods U_i of $W_i^u(K)'$ in $h_i'(K)$ and V_j of $W_j^s(K-1)$ in $h_j(K-1)$ such that if $x \in U_i$, $y \in V_j$, then $f_2(W_i^u(x)') \pitchfork W_j^s(y)$. Now, as in (B.1) and (B.3), we do a flattening isotopy with support disjoint from $\bigcup_{k < K} H(k)$, altering f_2 to f_3 so that if $x \in U_i$ and $f_2(x) \in W_j^s(K-1)$ then $f_2(W_j^u(x)') \supset V_j(f_2(x))$ where $V_j(y) = V_j \cap W_j^u(y)$. (We may shrink V_j slightly here.) The argument is similar to the proof of (B.3). We do both the perturbation of f_1 to f_2 and the isotopy to f_3 so that if $f_1(x) \in W_j^s(y)$, so is $f_3(x)$ (i.e., we flatten along stable disks). This is done so that when we repeat the argument on f^{-1} to get the analogous fitting for stable disks we will not destroy the unstable fittings. Now choose $t > 0$ sufficiently large that $g_t(h_i(K)) \cap h_i(K) \subset U_i$ for all i and $g_t(V_j(y)) \supset W_j^u(g_t(y))$ for all j and all $y \in W_j^s(K-1)$ and let $f_4 = g_t \circ f_3 \circ g_t$, then f_4 is the required map, i.e., if $x' \in h_i'(K)$, $y' = f_4(x') \in h_j(K-1)$ then $f_4(W^u(x')') \supset W_j^u(y')$.

Thus as remarked above if $f = f_4 \circ g_T$, and $x \in h_i(K)$, $y = f^n(x) \in h_j(K-1)$, then $f^n(W_i^u(x)) \supset W_j^u(y)$, and the same properties will hold for $g_t \circ f \circ g_{t'}$, $t, t' \geqslant 0$.

We now have f fitted for the K- and $(K-1)$-handles. We must repeat the entire argument to arrange that if $x \in h_i(K)$, $y = f^n(x) \in h_j(K-2)$, then $f^n(W_i^u(x)) \supset W_j^u(y)$. Because of the induction hypothesis and the fact that g_t is fitted no change at all is necessary if the orbit segment from x to y intersects $H(K-1)$, i.e., if $f^m(x) \in H(K-1)$ for some $m \leqslant n$. Thus in fitting f with respect to $H(K)$ and $H(K-2)$ we do not disturb the previously arranged fittings. We then proceed to the $(K-3)$-handles, etc., concluding the proof of the induction hypothesis for K.

Finally it is necessary to repeat the entire argument on f^{-1} to get $f^{-n}(W_i^s(x)) \supset W_j^s(y)$ whenever $x \in h_i(k)$, $y \in h_j(l)$, $x = f^n(y)$, $l \geqslant k$. The procedure we used was arranged so that this does not destroy the unstable fitting.

Once again, by choosing g_t with a rest point of index k at the barycenter of each k-simplex of some fine triangulation we can arrange that g_t is C^0 close to the identity for all t. Since all our deformations were small or involved composing with g_t, it follows that the fitted diffeomorphism f is C^0 close to the original map f_0. Q.E.D.

(B.5) REMARK. If f_0 is not fitted but for each k satisfies

(a) $f_0(M_k) \subset \text{int}(M_{k-1} \cup H(k))$, and

(b) $f_0(W_j^u(k)) \pitchfork W_i^s(k)$ for every pair of k-handles $h_i(k)$ and $h_j(k)$,

then the geometric intersection matrix $G(k)$ can be defined by $G(k)_{ij} =$ the number of points of intersection of $f_0(W_j^u(k))$ with $W_i^s(k)$. We can also define the algebraic intersection matrix $A(k)$ by $A(k)_{ij} = \Sigma \Delta_{ij}(x)$, $x \in f_0(W_j^u(k)) \cap W_i^s(k)$ where $\Delta_{ij}(x) = 1$ if the orientation of $f_0(W_j^u(k))$ and the orientation of the normal bundle of $W_i^s(k)$ (given by the orientation of $W_i^u(k)$) agree at x and $\Delta_{ij}(x) = -1$ otherwise. Moreover, examining the proof of (4.6) and (B.1) shows that $A(k)$ and $G(k)$ remain unchanged as f_0 is isotoped to a fitted diffeomorphism f.

(B.6) PROPOSITION [SS]. *Suppose $f_0: M \to M$ is a fitted diffeomorphism of a simply connected manifold of dimension $n \geqslant 5$. If there are no one-handles or $(n-1)$-handles in the handle sets $\{H(k)\}$ with respect to which f_0 is fitted then f_0 is isotopic to f_1 such that*

(a) *f_1 is fitted with respect to $\{H(k)\}$.*

(b) *The algebraic intersection matrix $A(k)$ for f_1 is the same as that for f_0.*

(c) *The geometric intersection matrix $G(k)$ for f_1 equals $|A(k)|$, i.e., $G(k)_{ij} = |A(k)_{ij}|$ for all i, j.*

PROOF. If the geometric intersection matrix $G_0(k)$ for f_0 has an entry $g_{ij} \neq \pm a_{ij}$, where a_{ij} is the ijth entry of $A(k)$, then there exist points $x, y \in f_0(W_j^u(k)) \cap W_i^s(k)$ with $\Delta_{ij}(x) = -\Delta_{ij}(y)$. That is $f_0(W_j^u(k))$ and $W_i^s(k)$ intersect with opposite intersection numbers at x and y.

The idea is to then use the so-called "Whitney lemma" to alter f_0 by an isotopy removing these two points of intersection without making any other changes in any geometric or algebraic intersection matrices. Notice that the effect of this change will be to leave

$A(k)$ unchanged but decrease g_{ij} by 2. Repeated application will result in $g_{ij} = |a_{ij}|$. Details of the "Whitney lemma" and its proof can be found in (6.6) of [M1]; we give only a sketch of the ideas involved.

The procedure is to form a closed loop consisting of an arc in $W_i^s(k)$ connecting x to y and an arc in $f_0(W_j^u(k))$ connecting x to y. Since M is simply connected this loop bounds a disk D (in fact an embedded disk since dim $M \geqslant 5$). By isotoping f_0, in a neighborhood of this disk, we can remove the two points of intersection (see Figure (B.3)).

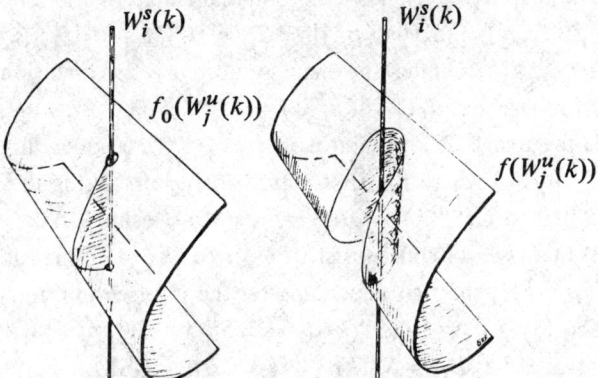

FIGURE (B.3)

In order that we do not create new points of intersection of stable and unstable manifolds and alter other intersection matrices it is important that int $D \cap W_i^s(j) = \emptyset$ for $j \geqslant 2$ and all i, and likewise that int $D \cap f_0(W_i^u(j)) = \emptyset$ for $j \leqslant n - 2$ and all i. These two conditions are completely dual, so we show only the first: int $D \cap W_i^s(j) = \emptyset$ for $j \geqslant 2$ and all i. If $j \neq 2$ we can by transversality perturb D to obtain disjointness.

Thus we need only show how int D can be made disjoint from $W_i^s(2)$. Take a loop γ' in D parallel to $\partial D = \gamma$ and close enough to it that there are no points in D between γ and γ' on stable manifolds of dimension $< n$. Using a gradient-like flow g_t, flow γ' down into a neighborhood of the one critical point p^0 of index 0 (there can be only one since there are no 1-handles). In this neighborhood it bounds a disk D' which is disjoint from $\{W_l^s(j)\}$ for all l and all $j \geqslant 2$. Thus flowing D' back up so its boundary coincides with γ' gives the desired D which is disjoint from $\{W_i^s(j)\}$.

Each time an alteration is done to decrease g_{ij} we can return to a fitted diffeomorphism without altering $\{A(k)\}$ or the other geometric intersection matrices (cf. (B.5)). Eventually we achieve $G(k) = |A(k)|$ for all k.   Q.E.D.

In Chapter 2 we discussed the construction of a chain complex C for M from a self-indexing Morse function $\phi: M \longrightarrow R$ and a gradient-like flow. Recall from (2.11) that if $M_k = \phi^{-1}((-\infty, k + 1 - \epsilon])$ then $C_k = H_k(M_k, M_{k-1})$ and $d_k: C_k \longrightarrow C_{k-1}$ is the boundary map in the long exact sequence of the triple (M_k, M_{k-1}, M_{k-2}). We also showed that if $\{p_i^k\}$ are the critical points of index k and $W^u(p_i^k)$ are assigned orientations then $\{[W^u(p_i^k)]\}$ form a basis of C_k, where $[W^u(p_i^k)]$ is the image of the generator of $H_k(W^u(p_i^k), W^u(p_i^k) \cap M_{k-1})$ under the inclusion $i: W^u(p_i^k) \longrightarrow M_k$. We want now to describe $d_k: C_k \longrightarrow C_{k-1}$ in terms of this basis. Let $S_i^u(k)$ denote the boundary of the k-disk $W_\epsilon^u(p_i^k)$, i.e., in local coordinates $(\vec{x}, \vec{y})$ centered at p_i^k, $S_i^u(k) = \{(\vec{x}, \vec{y}) | |\vec{x}|^2 = \epsilon, \vec{y} = 0\}$, so it is a $k-1$ dimensional sphere embedded in ∂M_{k-1}, and in fact equals $\partial M_{k-1} \cap W^u(p_i^k)$. The sphere $S_j^u(k)$ inherits an orientation from an orientation of $W^u(p_j^k)$.

(B.7) LEMMA. *If the gradient-like flow satisfies the transversality condition then* $d_k: C_k \longrightarrow C_{k-1}$ *has a matrix* D *with respect to the bases* $\{[W^u(p_j^k)]\}$ *and* $\{[W^u(p_i^{k-1})]\}$ *given by* $D_{ij} = S_j^u(k) \cdot W_i^s(k-1)$, *the intersection number of* $S_j^u(k)$ *and* $W_i^s(k-1)$.

PROOF. The map d_k is given by the boundary map of the long exact sequence of the triple (M_k, M_{k-1}, M_{k-2}), and $[W^u(p_j^k)] \in C_k = H_k(M_k, M_{k-1})$ is the image of a generator of $H_k(W_\epsilon^u(p_j^k), S_j^u(k))$ under the map induced by the inclusion $(W_\epsilon^u(p_j^k), S_j^u(k)) \longrightarrow (M_k, M_{k-1})$. It follows that $d_k([W^u(p_j^k)]) = [S_j^u(k)] \in C_{k-1} = H_{k-1}(M_{k-1}, M_{k-2})$ where $[S_j^u(k)]$ is the image of a generator of $H_{k-1}(S_j^u(k))$ under the composite $S_j^u(k) \longrightarrow M_{k-1} \longrightarrow (M_{k-1}, M_{k-2})$ where the first map is inclusion. Using the gradient-like flow g_t we can isotope $S_j^u(k)$ to $g_t(S_j^u(k)) \subset H(k-1) \cup M_{k-2}$ (simply take t large). If t is large enough $g_t(S_j^u(k)) \cap h_i(k-1)$ will consist of a set of $(k-1)$ dimensional disks each homologous to $\pm [W^u(p_i^{k-1})]$, the sign depending on the intersection number of the disk with $W^s(p_i^{k-1})$. Hence $[S_j^u(k)] = [g_t(S_j^u(k))] = \Sigma_i S_j^u(k) \cdot W^s(p_i^{k-1})[W^u(p_i^{k-1})]$, and it follows that the matrix D has $D_{ij} = S_j^u(k) \cdot W^s(p_i^{k-1})$.   Q.E.D.

We want now to investigate which chain maps $\{\alpha_k: C_k \longrightarrow C_k\}$ can be realized by fitted diffeomorphisms isotopic to a given one f_0. That is, can we find a diffeomorphism $f: M \longrightarrow M$ fitted with respect to the same handle sets as f_0 which induces a given chain map $\{\beta_k: C_k \longrightarrow C_k\}$ (i.e., $\beta_k = f_*: H_k(M_k, M_{k-1}) \longrightarrow H_k(M_k, M_{k-1})$). Since by (4.6) the matrix of α_k with respect to the basis $\{[W_i^u(k)]\}$ is the algebraic intersection matrix this is equivalent to asking what algebraic intersection matrices can be realized by an isotopic diffeomorphism fitted with respect to the same handle sets.

There are two conditions on $\{\beta_k\}$ which are obviously necessary and they turn out to be sufficient as well. First the chain map $\{\beta_k\}$ must be chain homotopic to $\{\alpha_k\}$ if it arises from a homotopic map. Secondly the matrix $B(0)$ for β_0 with respect to the basis $\{[W_i^u(0)]\}$ must have at most one nonzero entry in each column and it must be ± 1. This is because it is not possible to map a single 0-handle into more than one of the 0-handles since each 0-handle is a component of $H(0)$. Similarly if $B(m)$ is the matrix of β_m, $m = \dim M$, then each row must have at most one nonzero entry and this entry must be ± 1. This is the dual of the condition on $B(0)$. It says it is not possible to map parts of more than one m-handle to the same m-handle, which is clear for a fitted diffeomorphism.

(B.8) DEFINITION. If a chain map $\{\beta_k: C_k \longrightarrow C_k\}$ has β_m and β_0 represented by matrices $B(m)$ and $B(0)$ with respect to the bases given by m-handles and 0-handles, and $B(m)$ and $B(0)$ satisfy the conditions above, then we will say $\{\beta_k\}$ is a *geometric chain map*.

Note that the following result (in contrast to (B.6)) makes no assumptions about the dimension or simple connectivity of M.

(B.9) PROPOSITION [SS]. *Suppose* $f_0: M \longrightarrow M$ *is a fitted diffeomorphism of a connected manifold, and* $\{\alpha_k: C_k \longrightarrow C_k\}$ *is the chain map induced by* $f_{0*}: C_k \longrightarrow C_k$. *If* $\{\beta_k: C_k \longrightarrow C_k\}$ *is a geometric chain map, chain homotopic to* $\{\alpha_k\}$, *then* f_0 *is isotopic to* *to a diffeomorphism f_1 which is fitted with respect to the same handle sets as f_0 and has $\beta_k = f_{1*}: C_k \rightarrow C_k$ for each k.*

PROOF. Suppose that $\{s_k: C_k \rightarrow C_{k+1}\}$ is a chain homotopy from α to β, i.e., $d_{k+1} \circ s_k + s_{k-1} \circ d_k = \beta_k - \alpha_k$. We will isotope f_0 to achieve the effect of each s, one at a time. That is, we want to change α_k and α_{k+1} to $\alpha'_k = \alpha_k + d_{k+1} \circ s_k$ and $\alpha'_{k+1} = \alpha_{k+1} + s_k \circ d_{k+1}$ respectively, without altering α_j, $j \neq k$, $k + 1$.

Suppose at first that $k \neq 0$, 1 or $m - 1$. Now $s_k: C_k \rightarrow C_{k+1}$ can be written as $s_k = \Sigma a_{ij} t_{ij}$ where $a_{ij} \in Z$ and $t_{ij}: C_k \rightarrow C_{k+1}$ is given by $t_{ij}([W_j^u(k)]) = [W_i^u(k+1)]$ and $t_{ij}([W_l^u(k)]) = 0$ if $l \neq k$. Clearly it will suffice (by repeated application) if we do the case $s_k = \pm t_{ij}$.

To do this we first choose an arc $\gamma(t)$ embedded in M_{k+1} containing a point $x \in f_0(W_j^u(k)) \cap M_{k-1}$ and a point $y \in W_i^s(k+1)$. By transversality we can assume that except for x and y, γ is disjoint from stable and unstable manifolds of dimension $\leq m - 2$. These are stable and unstable manifolds with respect to the gradient-like flow g_t. In fact we can choose a tubular neighborhood U of γ satisfying

(a) U is diffeomorphic to $(0, 3) \times D^{m-1}$ and γ corresponds to $[1, 2] \times \{0\}$ (so x and y correspond to $(1, 0)$ and $(2, 0)$ respectively).

(b) Under this diffeomorphism $f_0(W_j^u(k)) \cap U$ corresponds to $\{1\} \times D^k \subset (0, 3) \times D^{m-1}$ and $W_i^s(k+1) \cap U$ corresponds to $\{2\} \times D^{m-k-1} \subset (0, 3) \times D^{m-1}$, while U is disjoint from all other stable manifolds of dimension $\leq m - 2$, and the image under f_0 of all other unstable manifolds of dimension $\leq m - 2$.

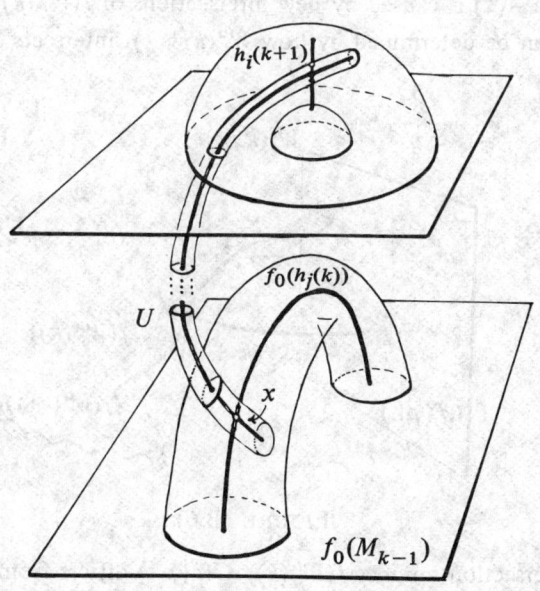

FIGURE (B.4)

We choose an isotopy ρ_t of $(0, 3) \times D^{m-1}$ such that

(a) Near the boundary of $[0, 3] \times D^{m-1}$, ρ_t is the identity for all $t \in [0, 1]$ and $\rho_0 = $ id on all of $[0, 3] \times D^m$.

(b) $\rho_1(\{1\} \times D^k)$ is wrapped around $\{2\} \times D^{m-k-1}$ (see Figure (B.5)). It can be linked in either direction around $\{2\} \times D^{m-k-1}$.

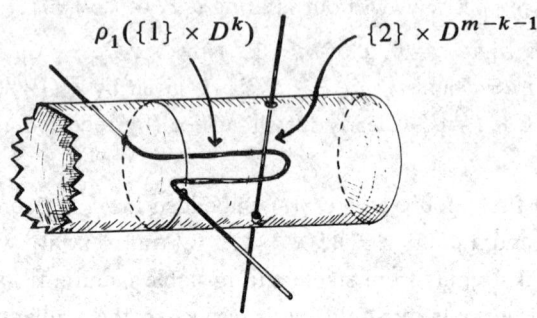

$\rho_1(\{1\} \times D^k)$ $\{2\} \times D^{m-k-1}$

FIGURE (B.5)

We now carry this isotopy over to $\hat{\rho}_t$ on M where it is supported in U and look at the new map $f = \hat{\rho}_1 \circ f_0$ and how the algebraic intersection matrices have changed. Clearly $A(l)$ has not changed if $l \neq k, k + 1$ since the change was supported on U which is disjoint from all stable and unstable manifolds not of dimension $k, k + 1, m - 1$, or m and our temporary assumption that $k \neq 0, 1, m - 1$ implies $m \geqslant 3$ so we are making no change in $A(m - 1)$ or $A(m)$. The change in $A(k)$ is caused by new intersections of $f(W_j^u(k))$ with $W^s(p_l^k)$ which occur in U. These can be determined by how $W_i^u(k + 1)$ intersects $W^s(p_l^k)$ (see Figure B.6).

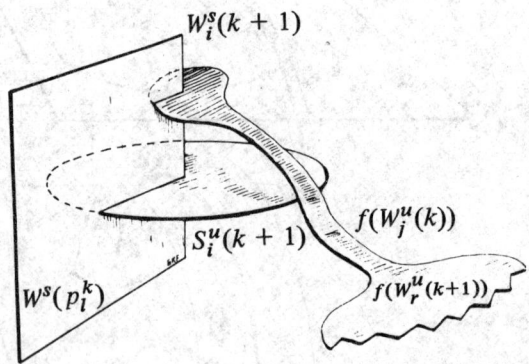

$W_i^s(k + 1)$

$S_i^u(k + 1)$ $f(W_j^u(k))$

$W^s(p_l^k)$ $f(W_r^u(k+1))$

FIGURE (B.6)

In fact the intersection number $f(W_j^u(k)) \cdot W^s(p_l^k)$ differs from $f_0(W_j^u(k)) \cdot W^s(p_l^k)$ by $\pm S_i^u(k + 1) \cdot W^s(p_l^k)$, the sign depending on the choice made for the direction $f(W_j^u(k))$ goes around $W_i^s(k + 1)$. (This portion of $f(W_j^u(k))$ can be isotoped onto most of $S_i^u(k + 1)$ in a way that includes all points of intersection either of them have with $W^s(p_l^k)$.)

Since by (B.7) the boundary map $d_{k+1}: C_{k+1} \longrightarrow C_k$ satisfies $d_{k+1}([W_i^u(k+1)]) = \Sigma_l S_i^u(k+1) \cdot W^s(p_l^k)[W_l^u(k)]$, it follows that the change in α_k is just $d_{k+1} \circ (\pm t_{ij})$ as desired. We want now to show the change in α_{k+1} is $(\pm t_{ij}) \circ d_{k+1}$. But this is also clear since the intersection number $f(W_r^u(k+1)) \cdot W^s(p_i^{k+1})$ differs from $f_0(W_r^u(k+1)) \cdot W^s(p_i^{k+1})$ by the coefficient of $[W_j^u(k)]$ in $d_{k+1}([W_r^u(k+1)])$, i.e., by the algebraic number of times $W_r^u(k+1)$ "bounds" on $W_j^u(k)$ (see Figure B.6). Thus this difference is $\pm t_{ij} \circ d_{k+1}$ as desired. By (B.5) we can isotope f to a fitted diffeomorphism without further altering the algebraic intersection matrices.

The case that $k = 1$ is essentially the same; the only difference being that it may not be possible for γ and U to miss $\{W_l^s(p_l^1)\}$ since these have dimension $m - 1$. Nevertheless, the intersection of $f(W_j^u(1))$ with $W^s(p_i^1)$ will consist of pairs of points with opposite intersection number (see Figure B.5) so the change in $A(1)$ and $A(2)$ will be as described above.

The case $k = 0$ is slightly different because of the necessity to keep the chain maps geometric at each stage. The chain homotopy $s_0: C_0 \longrightarrow C_1$ can be written as Σt_i where $t_i([W_i^u(0)]) = s_0([W_i^u(0)])$ and $t_i([W_j^u(0)]) = 0$ if $i \neq j$. Again, by repeated application, it will be sufficient to consider the case $s_0 = t_i$. Since both α_0 and $\alpha_0' = \alpha_0 + d_1 \circ t_i$ agree except on $[W_i^u(0)]$ and $\alpha_0'([W_i^u(0)]) = \beta_0([W_i^u(0)])$, it follows that the matrix $A(0)'$ for α_0 has at most one nonzero entry in each column and it is ± 1. Suppose $A(0)_{ji} = 1$ and $A'(0)_{ki} = 1$ (the case with minus signs is similar). Then $t_i([W_i^u(0)]) = c \in C_1$ with $d_1(c) = [W_k^u(0)] - [W_j^u(0)]$. We choose an embedded path γ in M_1 representing the chain c and with endpoints $W_j^u(0)$ and $W_k^u(0)$ but missing $W_l^u(0)$ if $l \neq j, k$. Initially we have $f_0(h_i(0)) \subset h_j(0)$. The isotopy to f must move $f_0(h_i(0))$ along in a tubular neighborhood of γ so that the new map f will have $f(h_i(0)) \subset h_k(0)$. The new algebraic intersection matrix on 0-handles will be $A(0)'$ and on one-handles will be the matrix for $\alpha + t_i \circ d_1$. We isotope so f is again fitted and repeat the process for the other t_j's.

The case $k = m - 1$ is completely dual to the case $k = 1$ and can be done by applying the same argument to f^{-1}. (Notice that for f^{-1}, k-handles become $(m - k)$-handles and the algebraic intersection matrices satisfy $A(k, f) = A^t(m - k, f^{-1})$ (cf. (3.3)).)   Q.E.D.

(B.10) PROPOSITION (SMALE). *Suppose M^n is connected, $\phi_0: M \longrightarrow R$ is a self-indexing Morse function and g_t' is a gradient-like flow with respect to ϕ_0. If $M_k = \phi_0^{-1}((-\infty, k + 1 - \epsilon]), 2 \leq k \leq n - 2$, and $C_k = H_k(M_k, M_{k-1})$ and a basis for C_k is given, then there is a new self-indexing Morse function $\phi: M \longrightarrow R$ and gradient-like flow g_t such that*

(a) *Outside of $M_k - M_{k-1}$, ϕ_0 and ϕ agree; on M_{k-1}, $g_t = g_t'$, $t \geq 0$, and on $M - M_k$, $g_{-t} = g_{-t}'$, $t \geq 0$,*

(b) *ϕ and ϕ_0 have the same critical points, and*

(c) *if $W^u(p_i^k)$ is the unstable manifold with respect to g_t, then $\{[W^u(p_i^k)]\}$ is the given basis of C_k.*

A proof of this result is given in (7.6) of [M1].

(B.11) PROOF OF (4.7). This now follows easily. The hypothesis of (4.7) says that $H_*(M)$ is isomorphic to $H_*(C)$. By (2.16) there is a self-indexing Morse function $\phi: M \longrightarrow R$

such that if $M_k = \phi^{-1}((-\infty, \; k + \frac{1}{2}])$ then C is isomorphic to the chain complex $\{H_k(M_k, M_{k-1}), \partial_k\}$. Moreover by (B.10) we can also assume that $\{[W^u(p_i^k)]\}$ corresponds to the given basis of C_k. By (4.6) we can isotope f to a diffeomorphism f_0 which is fitted with respect to handle sets $\{H(k)\}$ corresponding to ϕ. If $\alpha = \{\alpha_k \colon C_k \longrightarrow C_k\}$ is the chain map on $C_k \cong H_k(M_k, M_{k-1})$ induced by f_0 then $\tau_* = \alpha_* \colon H_*(C) \longrightarrow H_*(C)$. Since by hypothesis $H_*(C)$ is free it follows that α and τ are chain homotopic (see [D1, VI, 10.13]). Thus we can apply (B.9) and isotope f_0 to a fitted diffeomorphism f_0' whose algebraic intersection matrices are $\{A(k)\}$. Finally applying (B.6) we isotope f_0' to f_1 whose algebraic intersection matrices are $\{A(k)\}$ and whose geometric intersection matrices are $G(k) = |A(k)|, \; 0 \leqslant k \leqslant n$.   Q.E.D.

(B.12) PROOF OF (4.8). This proof is similar to the proof of (B.9), but easier. We give only a sketch; details can be found in (3.2) of [F1]. It is sufficient, by repeated application, to treat only the case when $P = P(k)$ has a single nonzero entry, $P_{ij} = 1$. We choose an arc γ running from $x \in f_0(W_j^u(k)) \cap M_{k-1}$ to $y \in W_i^s(k)$. Using transversality the arc γ can be chosen so that it misses all stable manifolds of dimension $m - k$ except at y, where $m = \dim M$, and lies entirely in M_k. This is true since $M_k -$ {stable manifolds of dimension $\leqslant (m - 1)$} is connected because any point in it is on a gradient-flow orbit which goes to M_0 which is connected. We can by a similar argument also choose γ so it misses the image under f_0 of all unstable manifolds of dimension k.

We now do an isotopy supported in a tubular neighborhood U of γ to increase the number of points of intersection of $f_0(W_j^u(k))$ with $W_i^s(k)$ by 2 (see Figure (B.7)).

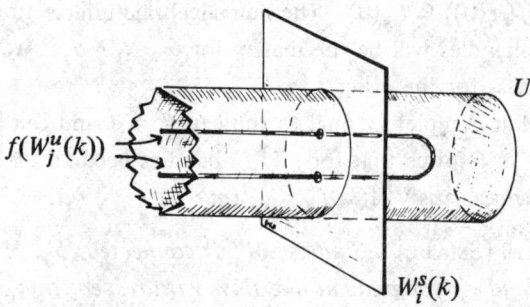

FIGURE (B.7)

If we then isotope so that we obtain a fitted diffeomorphism again (with respect to the same handle sets), the new geometric intersection matrices will be the same except the kth which will be $G(k) + 2P$, and the algebraic intersection matrices will be the same (cf. remark (B.5)).   Q.E.D.

References

[AM] M. Artin and B. Mazur, *On periodic points*, Ann. of Math. **81** (1969), 82–99.

[As1] D. Asimov, *Round handles and non-singular Morse-Smale flows*, Ann. of Math. **102** (1979), 41–54.

[As2] ——, *Homotopy of non-singular vector fields to structurally stable ones*, Ann. of Math. **102** (1979), 55–69.

[As3] ——, *Flaccidity of geometric index for non-singular vector fields*, Comment. Math. Helv. **52** (1977), 145–160.

[Bass1] H. Bass, *Introduction to some methods of algebraic K-theory*, C.B.M.S. Regional Conf. Series in Math., no. 20, Amer. Math. Soc., Providence, R. I., 1973.

[Bass2] ——, *The group SSF*, handwritten notes, 1979.

[Ba1] Steve Batterson, *Constructing Smale diffeomorphisms on compact surfaces*, Trans. Amer. Math. Soc. **257** (1980), 237–245.

[Ba2] ——, *The dynamics of Morse-Smale diffeomorphisms on the torus*, Trans. Amer. Math. Soc. **256** (1979), 395–403.

[Ba3] ——, *Orientation reversing Morse-Smale diffeomorphisms on the torus*, Trans. Amer. Math. Soc. (to appear).

[Bl-F] Paul Blanchard and John Franks, *The dynamical complexity of orientation reversing homeomorphisms of surfaces*, Invent. Math. **62** (1980), 333–339.

[Bl-F2] ——, *An obstruction to the existence of certain dynamics in surface diffeomorphisms*, Ergodic Theory Dynamical Systems **1** (1981), 255–260.

[BW] J. Birman and R. F. Williams, *Knotted periodic orbits in dynamical systems*. I, II (to appear).

[B1] R. Bowen, *Equilibrium states and the ergodic theory of Anosov diffeomorphisms*, Lecture Notes in Math., vol. 470, Springer-Verlag, Berlin and New York, 1975.

[B2] ——, *On Axiom A diffeomorphisms*, C.B.M.S. Regional Conf. Series in Math., no. 38, Amer. Math. Soc., Providence, R. I., 1978.

[B3] ——, *Entropy versus homology for certain diffeomorphisms*, Topology **13** (1974), 61–67.

[B4] ——, *One-dimensional hyperbolic sets for flows*, J. Differential Equations **12** (1972), 173–179.

[B5] ——, *Topological entropy and Axiom A*, Proc. Sympos. Pure. Math., vol. 14, Amer. Math. Soc., Providence, R. I., 1970, pp. 23–41.

[BF]   R. Bowen and J. Franks, *Homology for zero dimensional basic sets*, Ann. of Math. **106** (1977), 73–92.

[BL]   R. Bowen and O. Lanford, *Zeta functions of restrictions of the shift transformation*, Proc. Sympos. Pure Math., vol. 14, Amer. Math. Soc., Providence, R. I., 1970, pp. 43–49.

[C]   Charles Conley, *Isolated invariant sets and the Morse index*, C.B.M.S. Regional Conf. Series in Math., no. 38, Amer. Math. Soc., Providence, R. I., 1978.

[D1]   A. Dold, *Lectures on algebraic topology*, Springer-Verlag, Berlin, 1972.

[D2]   ———, *Fixed point index and fixed point theorem for Euclidean neighborhood retracts*, Topology **4** (1965), 1–8.

[F-S]   J. Franke and J. Selgrade, *Hyperbolicity and chain recurrence*, J. Differential Equations **26** (1977), 27–36.

[F1]   J. Franks, *Constructing structurally stable diffeomorphisms*, Ann. of Math. **105** (1977), 343–359.

[F2]   ———, *A reduced zeta function for diffeomorphisms*, Amer. J. Math. **100** (1978), 217–243.

[F3]   ———, *Morse inequalities for zeta functions*, Ann. of Math. **102** (1975), 143–157.

[F4]   ———, *The periodic structure of non-singular Morse-Smale flows*, Comment. Math. Helv. (1978), 279–294.

[F5]   ———, *Morse-Smale flows and homotopy theory*, Topology **18** (1979), 199–215.

[F6]   ———, *Knots, links and symbolic dynamics*, Ann. of Math. **113** (1981), 529–552.

[F7]   ———, *Non-singular flows on S^3 with hyperbolic chain recurrent set*, Rocky Mountain J. Math. **7** (1977), 539–546.

[FN]   J. Franks and C. Narasimhan, *The periodic behavior of Morse-Smale diffeomorphisms*, Invent. Math. **48** (1978), 279–292.

[F-Sh]   J. Franks and M. Shub, *The existence of Morse-Smale diffeomorphisms*, Topology **20** (1981), 273–290.

[Fr1]   D. Fried, *Cross-sections to flows*, Ph.D. thesis, University of California, Berkeley, 1976.

[Fr2]   ———, *Flow equivalence, hyperbolic systems, and a new zeta function for flows* (to appear).

[Fr3]   ———, *Subshifts on surfaces*, preprint.

[G]   F. R. Gantmacher, *Theory of matrices*. II, Chelsea, New York, 1960.

[Gu]   John Guckenheimer, *Bifurcations of dynamical systems*, Dynamical Systems, C.I.M.E. Lectures (Bressanone, Italy, 1978), Birkhauser, Basel, 1980, pp. 115–232.

[H]   Mike Handel, *The entropy of orientation reversing homeomorphisms of surfaces*, preprint.

[HPPS]   M. Hirsch, J. Palis, C. Pugh and M. Shub, *Neighborhoods of hyperbolic sets*, Invent. Math. **9** (1970), 121–134.

[H-P] M. Hirsch and C. Pugh, Proc. Sympos. Pure Math., vol. 14, Amer. Math. Soc., Providence, R.I., 1970, pp. 133–163.

[L] S. Lang, *Algebra*, Addison-Wesley, Reading, Mass., 1965.

[Ma] R. Mañe, untitled preprint.

[Mal] M. Maller, *Fitted diffeomorphisms of non-simply connected manifolds*, Topology **19** (1980), 395–410.

[Mar] L. Markus, *Lectures in differentiable dynamics*, C.B.M.S. Regional Conf. Series in Math. vol. 3, Amer. Math. Soc., Providence, R. I., 1971, (revised 1980).

[M1] J. Milnor, *Lectures on the H-cobordism theorem*, Princeton Univ. Press, Princeton, N. J., 1965.

[M2] ———, *An introduction to algebraic K-theory*, Ann. of Math. Studies, no. 72, Princeton, N. J., 1971.

[M3] ———, *Infinite cyclic coverings*, Conf. Topology of Manifolds (Michigan State, 1967), Prindle, Weber and Schmidt, Boston, Mass., 1968, pp. 115–133.

[M4] ———, *Topology from the differentiable viewpoint*, University Press of Virginia, Charlottesville, Va., 1965.

[Mo] J. Morgan, *Non-singular Morse-Smale flows in 3-dimensional manifolds*, Topology **18** (1978), 41–53; Errata.

[N] Carolynn Narasimhan, *The periodic behavior of Morse-Smale diffeomorphisms on compact surfaces*, Trans. Amer. Math. Soc. **248** (1979), 145–169.

[Ne] M. Newman, *Integral matrices*, Academic Press, New York, 1972.

[New] S. Newhouse, *Lectures on dynamical systems*, Dynamical Systems, C.I.M.E. Lectures (Bressanone, Italy, 1978), Birkauser, Basel, 1980, pp. 1–114.

[P] M. Peixoto, *Structural stability on two-dimensional manifolds*, Topology **1** (1962), 101–120.

[Pa-S] W. Parry and D. Sullivan, *A topological invariant of flows on one-dimensional spaces*, Topology **14** (1975), 297–299.

[P-S] C. Pugh and M. Shub, *The Ω-stability theorem for flows*, Invent. Math. **11** (1970), 150–158.

[Ro] J. Robbin, *A structural stability theorem*, Ann. of Math. **94** (1971), 447–493.

[R1] C. Robinson, *Structural stability of C^1 flows*, Dynamical Systems (Warwick, 1974), Lecture Notes in Math., vol. 468, Springer-Verlag, Berlin and New York, 1975, pp. 262–277.

[R2] ———, *Structural stability for C^1 diffeomorphisms*, J. Differential Equations **22** (1976), 28–73.

[S1] S. Smale, *Differentiable dynamical systems*, Bull. Amer. Math. Soc. **73** (1967), 797–817.

[S2] ———, *The Ω-stability theorem*, Proc. Sympos. Pure Math., vol. 14, Amer. Math. Soc., Providence, R. I., 1970, pp. 289–297.

[S3] ———, *On gradient dynamical systems*, Ann. of Math. **74** (1961), 199–206.

[S4] ———, *Diffeomorphisms with many periodic points*, Differential and Combinatorial Topology (S. Cairns, ed.) Princeton Univ. Press, Princeton, N. J., 1965.

[S5] S. Smale, *Notes on differentiable dynamical systems*, Proc. Sympos. Pure Math., vol. 14, Amer. Math. Soc., Providence, R. I., 1970, pp. 277–287.

[S6] ——, *Morse inequalities for a dynamical system*, Bull. Amer. Math. Soc. **66** (1960), 43–49.

[Sh] M. Shub, *Stabilité globale des systèmes dynamique*, Astérisque **56** (1978).

[SS] M. Shub and D. Sullivan, *Homology and dynamical systems*, Topology **14** (1975), 109–132.

[T] G. Torres, *On the Alexander polynomial*, Ann. of Math. **57** (1953), 57–89.

[W] W. Wilson, *Smoothing derivatives of functions and applications*, Trans. Amer. Math. Soc. **139** (1969), 413–428.

[Wm1] R. Williams, *Classification of subshifts of finite type*, Ann. of Math. **98** (1973), 120–153; Errata **99** (1979), 380–381.

[Wm2] ——, *The zeta function of an attractor*, Conf. Topology of Manifolds (Michigan State, 1967) Prindle, Weber and Schmidt, Boston, Mass., 1968.

[Wm3] ——, *The structure of Lorenz attractors*, Inst. Hautes Étude Sci. Publ. Math. **50** (1979), 73–100.

[Wh] J. H. C. Whitehead, *The mathematical works of J. H. C. Whitehead*, vol. 3, Pergamon Press, Oxford, 1962.

[Z] C. Zeeman, *Morse inequalities for diffeomorphisms with shoes and flows with solenoids*, Dynamical Systems (Warwick, 1974), Lecture Notes in Math., vol. 468, Springer-Verlag, Berlin and New York, 1975, pp. 44–47.

ABCDEFGHIJ–AMS–898765432